Desert Energy

Desert Energy examines the key technologies being deployed in an effort to tap the potential presented by the world's deserts for siting large-scale solar power applications, and surveys the feasibility of such projects given the remoteness and hostility of these environments. Focusing on large-scale photovoltaics and concentrating solar thermal power, the book explains how the systems work, projects that are being planned, the required scales, and the technical difficulties they need to overcome to function effectively. It then moves on to examine the economics of such projects (including financing) and the social and environmental effects they may have. The book also considers the future for these systems as well as other, less developed technologies which may have a role to play. With reference throughout to built or planned projects, and written in a clear, jargon-free style, this is a must-read for anyone interested in the development of large-scale solar applications.

Alasdair Cameron is a UK-based writer and campaigner on environmental issues. He has written extensively on renewable energy and was formerly Assistant Editor of *Renewable Energy World* magazine. He currently lives in London.

Desert Energy

A guide to the technology, impacts and opportunities

Alasdair Cameron

LIBRARY OF
CONGRESS
WITHDRAWN
ADDITIONAL
SERVICE COPY

First edition published 2013
by Routledge
2 Park Square, Milton Park, Abingdon, Oxon, OX14 4RN

Simultaneously published in the USA and Canada
by Routledge
711 Third Avenue, New York, NY 10017

Routledge is an imprint of the Taylor & Francis Group, an informa business

© 2013 Alasdair Cameron

The right of Alasdair Cameron to be identified as author of this work has been asserted by him in accordance with sections 77 and 78 of the Copyright, Designs and Patents Act 1988.

All rights reserved. No part of this book may be reprinted or reproduced or utilised in any form or by any electronic, mechanical, or other means, now known or hereafter invented, including photocopying and recording, or in any information storage or retrieval system, without permission in writing from the publishers.

British Library Cataloguing in Publication Data
A catalogue record for this book is available from the British Library

Library of Congress Cataloging-in-Publication Data
Cameron, Alasdair, 1981–
Desert energy: a guide to the technology, impacts, and opportunities / Alasdair Cameron. – 1st ed.
    p. cm.
  1. Solar power plants. 2. Solar thermal energy. 3. Arid regions. I. Title.
  TK1056.C36 2012
  621.31´244–dc23   2011047657

ISBN: 978-1-84971-184-5 (hbk)
ISBN: 978-0-203-11841-2 (ebk)

Typeset in Sabon
by HWA Text and Data Management, London

Printed and bound in Great Britain by the MPG Books Group

Contents

List of figures	vi
List of tables	vii
Foreword	viii
Acknowledgements	x
1 Introduction	1
2 Technology	20
3 Energy from the desert	59
4 Economic and policy aspects of solar power, and the status of regional markets	69
5 Existing and planned projects	96
6 Long-term visions	120
7 Environmental and resource issues facing solar technology	142
8 Conclusion	159
Notes	162
Index	176

Figures

1.1	Domestic solar system on the author's house in London, using monocrystalline	2
2.1	Parabolic trough plant in the US	31
2.2	LS-2 collector with torque tube structure	33
2.3	Truss support structure from LS-3 parabolic trough collector	34
2.4	The space frame structure used by Solargenix	35
2.5	Experimental solar power tower in California	38
2.6	Dish–Stirling system in Maricopa	40
2.7	Linear Fresnel reflectors combine attributes of power towers and parabolic troughs	42
2.8	A prototype solar chimney or updraft tower in Spain	43
2.9	Crystalline PV in use in Villar de Cañas in Spain	46
2.10	Thin-film CdTe photovoltaics in use in Canada	49
2.11	Concentrating photovoltaic systems in Puertollano, Spain	52
3.1	Satellite image of Siwa oasis in Egypt, 2009	60
4.1	Small-scale photovoltaic installation in Morocco	91
5.1	SEGS IV parabolic trough power plant in California, USA	100
5.2	The Solucar complex in Spain includes the PS10 and PS20 power towers	110
5.3	The 30 MW cadmium telluride Cimarron PV plant in New Mexico, USA	117
6.1	Artist's impression of aerial view of proposed master plan of Masdar City (southern orientation)	123
6.2	Predicted costs of renewables under the CSP-Med scenario	129

Tables

1.1	Total and per capita greenhouse gas emissions from energy consumption in 2009	15
2.1	Reported and estimated performance of various CSP technologies	24
2.2	Water use of CSP technologies, including washing and cleaning	29
2.3	SEGS I–IX history and operational statistics (all in California)	32
2.4	Costs of HVAC and HVDC transmission	57
3.1	World deserts by size and insolation	63
4.1	Cost estimates for 100 MW PV installations	71
4.2	Feed-in tariffs for photovoltaics and CSP in selected countries	77
4.3	PV installation rates in key markets 2007–2010 (in MW)	85
4.4	Renewable Portfolio Standard by US state	87
4.5	Feed-in tariffs in Australia	95
5.1	Large CSP installations in the USA	99
5.2	Large CSP installations under development in the USA	102
5.3	CSP projects operational in Spain, end 2010	106
5.4	List of CSP projects reportedly under construction or in pre-construction in Spain, March 2010	108
5.5	Largest PV systems in the world, February 2011	113
5.6	Large-scale PV plants completed or planned in the USA	116
5.7	Large-scale PV plants in China	118
6.1	Phases and targets of Jawaharlal Nehru National Solar Mission 2010–2022	138
7.1	Water use of conventional and CSP technologies, including washing and cleaning	149

Foreword

In 2004, the German Aerospace Center (DLR) got an order from the German Federal Ministry for the Environment, Nature Conservation and Nuclear Safety (BMU) to assess the feasibility of a fascinating new idea to use the deserts south of the Mediterranean Sea for sustainable power generation, not only for the region itself, but also for Europe. The task was not easy and took three years and three studies to be solved. First of all, we had to define the term 'sustainable' in order to produce a reasonable target function for our analysis. Then we had to assess the demand expectations of each of the 50 countries north and south of the Mediterranean, we had to quantify the renewable and the conventional potential for power generation and the related development of the electricity cost, and finally we had to assess the socio-economic and environmental impacts of a transition towards such a new energy supply scheme. What we found was rather surprising, at least from the point of view of the year 2005:

1 Sustainable power supply must be affordable, secure and compatible without compromise.
2 Electricity demand growth rates increase towards the South, with the highest increase of electricity demand in the coming 40 years taking place south of the Mediterranean Sea.
3 The potential to produce renewable electricity, particularly solar, is several orders of magnitude larger than demand will ever be, in spite of excluding all areas that would not be suitable, like agricultural land, forests, sand dunes, steep mountain slopes or national parks.
4 A future electricity mix based mainly on renewable energy will not create a considerable economic burden, but will relieve customers and governments from increasing fossil fuel prices and energy subsidies.
5 The international goals for reducing carbon emissions can be achieved in spite of a strong economic development and growth of population south of the Mediterranean.
6 A future sustainable electricity mix will have to be based on roughly 50% variable renewables like wind power and photovoltaics and on

50% flexible renewables capable of producing electricity on demand, like hydropower, biomass, geothermal energy, and concentrating solar thermal power from deserts. Until this is fully achieved, flexible power will have to be provided by fossil fuels.
7 The main barrier to overcome is a change of political and economic structures and thinking.

Five years after finishing our studies we are seeing a very dynamic development of renewable energy in EUMENA (Europe, Middle East & North Africa – a term created by Dr Gerhard Knies during the studies) leaving behind even our most optimistic expectations, and a lot of interest in renewables and Desert Energy. And, we also see a lot of new thinking.

The book at hand is a very good example of this new thinking, it gives an understandable overview of the chances and challenges involved, on lessons learned and on some questions that still have to be answered in the future. The author is neither too optimistic nor too pessimistic about Desert Energy, but simply realistic, and in this way has managed to write an easily understandable, comprehensive book about the Desert Energy concept which is highly recommendable for everybody interested in the topic.

Franz Trieb
DLR Stuttgart
June 2012

Acknowledgements

I would like to thank all the people who helped me with this book, whether through correcting errors, suggesting points for discussion or simply listening to me endlessly elaborating on the potential for solar power to transform our lives. Specific thanks need to go to David Faiman, Paul Maycock and David Cameron for their input into the early drafts, while gratitude is due to the endlessly patient Kat Hollaway and her colleagues at Taylor and Francis. Special thanks also need to go to Michael Fell, formerly of Earthscan Publishing, who first suggested the idea of writing this book, and to Yasmeen Ismail, for allowing me to spend so many weekends looking at tables and print-outs of costs-per-watt.

Writing a publication like this is a process of exploration, and I am painfully aware of how incomplete and superficial its treatment of many of the topics necessarily is. Every paragraph of this volume could probably be expanded into a book in its own right, and I suspect many have. I am also in full knowledge of the fact that in a subject as fast moving as this, many individual projects or economic measures will go quickly out of date. Governments can change, companies can go bust and promising technologies can disappear entirely. None the less, I hope it will prove a useful and interesting introduction to solar power and encourage the reader to explore further, giving them the tools to understand the latest developments in what will almost certainly be one of the biggest technological stories of the twenty-first century.

Chapter 1
Introduction

This book is about large-scale solar energy from the desert – what it is, why we need it and what is actually happening in this field. The idea is to shed some light on one of the most exciting areas of renewable energy generation by providing an introduction to the various solar technologies which may be used for large-scale electrical generation, how they are being deployed right now and what might happen in the future. It will also review the policy environment in which these technologies are being developed and what barriers, technical, financial and environmental may need to be overcome. Solar energy has the potential to revolutionise our economies and our lives, and make a real contribution to developing a truly sustainable and prosperous future. How do we get there? Rather than aiming at the specialist, this book hopes to be of use to those seeking a general understanding of the large-scale solar energy business, from a technical, financial and policy perspective.

Despite the fact that it has been around for more than half a century, there is still something about solar energy that seems space-age and futuristic. The idea that we can generate electricity from 'nothing' seems like something from a science fiction novel. Of course, as we shall see the idea of free energy is still a very long way off, but the sheer elegance of using sunlight to power our lives has a certain utopian appeal. Perhaps this is one of the reasons it has taken so long for it to be regarded seriously by the general population – it simply seems too good to be true. There are other reasons also. Despite rapidly falling costs, the technology is still relatively expensive. While the solar fuel may be free, the equipment to harness it and the grid lines to carry it are most certainly not. Nonetheless, when I finally got my own solar panels installed at the beginning of 2010, it felt like a revolution (Figure 1.1). It wasn't cheap, and even with the UK's new feed-in tariff it will take ten years to pay itself back, but as I sit here in my flat in London's Camden Town, with my laptop being powered by the late spring sunshine, the concept makes a great deal of sense.

2 Desert Energy

Figure 1.1 Domestic solar system on the author's house in London, using monocrystalline photovoltaic panels

Large solar or very large solar?

Perhaps one of the first things to look at is what is meant by large scale. Solar plants come in all shapes and sizes, and ten thousand small roof-top plants in a town or district may have the same generation potential as one vast array mounted on the plains. For the purposes of this book though, large-scale solar will be mostly restricted to those plans and technologies that can one day supply thousands of megawatts of utility scale electricity, from large installations. Of course there is a long way to go before plants in the multi-gigawatt class appear, so in practice centralised solar plants of more than 10 MW, or smaller demonstration and pre-commercial plants of promising technologies will be discussed. While these are still small compared with fossil powered thermal stations, most of which have capacities of well over 1000 MW, all of the technologies discussed in this book have been chosen because they have the potential to provide electricity on equivalent scales and so become leading sources of energy in the twenty-first century. It is therefore important to explore these and to look at how the industry is developing.

The timing of the book is no accident. Had it been written just a few years ago, most of the projects and possibilities would have been purely hypothetical. When I first started writing about renewable energy – just a few years ago in 2005 – the largest photovoltaic plant in the world was less than 10 MW. Now there are over a hundred meeting that benchmark, and the largest PV plants in development are several hundred megawatts apiece. Concentrating solar too is enjoying a renaissance, with dozens of plants under construction in California, Spain and elsewhere. In the last five years solar power has undergone a revolution in how it is perceived and in terms of the scale of what is being constructed. This is partly as a result of technological advances, but also because the global policy environment has shifted, with greater recognition of the environmental and energy problems that the world is facing, and the urgent need to act. Because of this, many countries and regions have adopted financial instruments to encourage the growth of low carbon or oil-free energy and investment in the sector has soared. New technologies have come on-stream and old ones have been brought back to life. There has also been a realisation among governments throughout Europe, North America and Asia that renewable energy may emerge as *the* crucial industry of the future, and many nations are beginning to link their hopes for economic recovery and renewed industrial power on getting an early advantage in these sectors. Some countries of course had this foresight many years ago. Germany and Japan were the two early movers in solar photovoltaics, followed by the USA and Spain. Between them these countries are now not only the largest markets for solar technologies, but are also world leaders in the design and manufacture of certain kinds of photovoltaic and concentrating solar power equipment, from solar cells to panels, control units, generators and mirrors. Germany in particular deserves credit for supporting the photovoltaic industry, with its dedicated feed-in tariff being largely responsible for the progression of the sector over the last 15 years (the fact that it has recently begun to cut its tariff is as much a sign of the strength of the market as anything else). China and Taiwan too have become major players, particularly in manufacturing, where they have taken a huge slice of the global silicon PV market in just a few years,[1] and undoubtedly new regions will enter the arena in the coming decades. In the concentrating solar sector, it was the United States that took the lead in the 1970s, investing in research and development in western desert states. More recently Spain has taken over the mantle, cashing in on its sunshine to become a world leader, with companies that are now set to export their skills all over the world.

The renewable imperative

Any look at the solar energy sector must look at the environment in which it exists. As I write, the need for practical solutions to combat the environmental and resource crises faced by human societies has never been greater. Indeed, as many have said, it will be the defining challenge of the twenty-first century. Sadly, our efforts, so far, are failing. Everywhere we look the environment can no longer take the strain that is already being placed upon it, and consumption and populations are rising. Fisheries are collapsing or in danger of collapsing,[2] threatening the foundations of our food supply. Biodiversity and abundance of wildlife is declining and biologists warn that we are in the midst of the earth's sixth great extinction event, this time initiated not by a geological or climatic upheaval, but by human actions.[3] Along with the extinctions, refuse is piling up. In the oceans it seems even the most remote beaches and currents are often clogged with plastic. On a recent visit to Ascension Island in the mid-Atlantic, I was saddened to see isolated beaches littered with plastic waste and solidified crude oil, despite the fact that the island is 1500 miles (2500 km) from the nearest mainland. Deforestation too continues apace, with local and global consequences. The forests of central Africa are home to our closest relatives, the gorillas and the chimpanzees, while at the same time the water vapour which gathers over the Congo basin generates rainfall not only across Africa, but as far away as the US Midwest and Eastern Europe,[4] supplying important food producing regions. Yet these vital forests are being logged at an extraordinary rate to feed the consumers of Asia, the Middle East and the West. If the forests of the Congo and West Africa disappear, so too might the great rivers of Africa, and the civilisations they nourish. Water itself is running scarce, with the World Health Organization estimating that 2.6 billion people suffer from inadequate sanitation, and one billion lack safe drinking water.[5] And the problem will only get worse. By 2050, the global population will have increased by an additional 40 per cent. As an example, the Indian government estimates that the vast nation, with its one billion plus people, will be facing water shortage by the middle of the century – and this before many of them even have mains water.[6]

Then there is climate change caused by the burning of fossil fuels, which remains a serious threat to our civilisations. In societies as finely balanced and interlinked as ours even modest changes in rainfall, ocean currents, diseases and agricultural productivity could have dramatic consequences. The recent eruptions of volcanoes in Iceland, while unconnected to climate change, show the disruption that can be caused by even modest changes to our infrastructure – and these lasted just a few days.[7] And there is no guarantee

---

Box 1.1 **What kind of solar?**

While people often use the term to refer solely to electricity producing technologies, strictly speaking solar power can refer to any mechanism for extracting useful energy from the sun. For the purposes of this book, solar technologies will be largely divided into two categories – technologies incorporating light-sensitive materials that 'use' the photons in sunlight to produce a charge (photovoltaics or PV), and those solar thermal technologies which focus sunlight to produce heat, which can be used to drive turbines and produce electricity (concentrating solar power or CSP).

that the changes we could see will be modest. The Intergovernmental Panel on Climate Change warns us that a mere 2 °C rise in temperature could place 30 per cent of species at risk of extinction,[8] and cause shortages of food and water.[9] It does not take much imagination to see how resource shortages and changing weather patterns could trigger massive movements of people leading to migration, conflict and famine. Sadly, it now seems increasingly unlikely that we will be able to keep global temperature rise below that 2 °C threshold, and the race is now on to ensure that we do not greatly exceed it, with truly severe consequences. We need to take urgent measures to slow down and reduce the extent of the climate crisis.

At the same time as we are facing overwhelming environmental challenges, billions of people still live in grinding poverty. By some estimates around 1.1 billion live in absolute poverty, defined as less than a dollar (US) a day in purchasing-power adjusted income.[10] While this is far from a perfect measure of poverty, it gives some idea of the scale of the situation. Although this number has decreased thanks to the economic development of Asia, it continues to increase in many areas due in part to rapid population growth. To improve the quality of life for these people, and to stabilise the human population, the world needs to find mechanisms to get education, energy and economic development to those in poverty, but in a way that does not further devastate the environment and endanger all our futures.

Against the scale of problems facing us it can be hard to be optimistic that a workable solution to the environmental crisis can be found, but this is sad because many of the technical tools are already available, and there are many who feel that the shift to a greener, more efficient, more environmentally respectful society will have positive benefits reaching every aspect of our lives. Better fishing practices for example, while causing some short-term pain to fishing communities, could actually allow total catches to increase over time (the UK catch for example has fallen fourfold since 1900, despite vast improvements in technology).[11] Renewable energy, while not free of consequences, is a realistic possibility, and one that would bring untold benefits, not just reduced emissions. There was a cartoon by Joel Pett published in the magazine *USA Today* that showed a man at a conference displaying a slide of the diverse positive results that a move to a low-carbon economy would bring – energy independence, preserved rainforests, clean water, healthy air, green jobs, more liveable cities etc. In the background a man is standing up crying: 'But what if it's all a hoax and we create a better world for nothing?'[12] There is little evidence to suggest that climate change is a hoax, but the point still rings true.

The good news is that governments and societies have gradually begun to realise that they cannot continue as they have been, and slowly but surely structures are being put in place to try and deal with the environmental problems. From the much-hyped Kyoto Protocol or the Conference of Parties to the UN Framework Convention on Climate Change (UNFCCC), to the much derided Convention on Biological Diversity (CBD) the nations of the world are in conversation as never before on the issues of the environment. Individual regions and countries too have begun to take action, developing stricter emissions standards or introducing laws to encourage clean alternatives to fossil fuels. Communities, from India to Kenya to the USA and Europe, are seeking to change the way they do things, through campaigning for political change or through town-level initiatives. Civil disobedience, so-called direct action, has hit the headlines all over the world, from anti-coal-mining protests in India and the USA to the 'Climate Camp' protests that helped to prevent the speedy construction of a new coal power plant at Kingsnorth in the UK. So far these measures are nowhere near enough, and it remains to be seen whether such local and inter-governmental efforts will ultimately be successful in addressing our problems, but the existence of these movements is significant in itself. Human society is at a crossroads. The question of whether we can respond to the problems we face is still open, but the solutions are in our hands. The choices we make in the coming decades will decide on the outcome.

The climate crisis

The debate over whether human actions are causing climate change has been difficult and bitter, and far longer than many imagine. Indeed the concept was first put forward in the late nineteenth century by a number of researchers, such the Swedish scientist Svante Arrhenius, building on earlier work by Joseph Fourier and John Tyndall. Indeed it was John Tyndall (after whom the UK's Tyndall Centre for Climate Change Research is named) who is credited with first working out the greenhouse potential or radiative forcing of a number of gases, including water vapour and carbon dioxide.[13] At the time that these men were working it may never have occurred to them that humankind was capable of burning enough fossil fuels to alter the climate, such has been the expansion of human activity in the last 100 years. While there is still tremendous uncertainty over the course which climate change will take or the speed and scale of the change we can expect, there is a broad and multi-disciplined scientific consensus that the world is warming, and that human actions are responsible. Science does not deal in certainties, but probabilities, and all we can do is act on the best evidence available to us.

No one knows exactly what climate change could bring in the future. Current figures show that the Earth has already warmed by 0.6 ± 0.2 °C while mainstream estimates for the coming century suggest a range of warming from 1.1 to 6.4 °C, depending on a number of scenarios and human actions (crucially our efforts to reduce emissions).[14] Feedback mechanisms are particularly important. There is some evidence that the world possesses tipping points which, if breached, could lead to a positive warming feedback and runaway climate change. The disappearance of the Greenland ice-sheets or the warming of Siberia are examples of such possible feedbacks. Both the Greenland ice-sheet and the tundra of Eurasia appear white, reflecting a huge amount of the sun's energy back into space (this is known as the albedo effect). As they warm, the ice sheets melt and the tree line of Siberia and Canada moves further north. This reduces the amount of 'white space', causing the Earth to absorb more solar radiation. At the same time, the permafrost of the far north contains colossal quantities of methane, stored in the soil. As the ground gradually heats up and melts, this methane is released. Since methane is a powerful greenhouse gas in its own right (over twenty times more radiative forcing than carbon dioxide) this again can add to the problem, triggering more warming and more methane release.

Despite all the uncertainty, most predictions suggest that the consequences of climate change will be generally negative and in some cases could be devastating. Floods could inundate low-lying countries. Crop yields may increase in some places, but overall they can be expected to decline, placing millions at risk of starvation. Rainfall could decrease in the areas that need it most, like Sub-Saharan Africa and Southern Europe, while flooding will increase in others. The monsoon rains that supply the multitudes of Asia could become weaker or more erratic. Even tectonic activity could be affected. Some studies suggest changes in water density and ice sheets can shift pressures on the earth's crust, triggering earthquakes and seismic activity.

Importantly, it is not even necessary that the world becomes less habitable as a whole. What matters is that the areas currently most conducive to human civilisation and population may change, causing mass migration and civil conflict. Many people in both the rich and poor world are concerned by uncontrolled immigration, yet it is hard to imagine the scale of forced migration that could occur if swathes of India or Bangladesh were to become uninhabitable or Sub-Saharan Africa were to dry out (not to mention the possible flooding of the Netherlands or New York City). Hundreds of millions could be forced to move, and it is unlikely that they would be constrained by things such as borders or walls.

To have any certainty of avoiding a level of change much above 2 °C, scientific opinion estimates that we need to keep atmospheric carbon levels below 450 parts per million.[15,16,17] Recent readings from the high altitude Mauna Loa observatory in Hawaii show that atmospheric carbon is already approaching 393 ppm.[18] According to the IPCC we need urgent and deep cuts in greenhouse gas emissions from developed nations, up to 80–90 per cent by 2050, followed in the future by a slowing growth and eventual reduction in emissions from developing economies, to provide a reasonable chance that this level will reduce and stabilise, causing no more than 2 °C of warming, a level that will have serious consequences, but that may not be catastrophic. Since nearly 22 per cent of global greenhouse gas emissions come directly from electricity production (and this is higher in some rich countries, 34 per cent in the USA for example),[19] this is clearly an important part of the problem, and a good place to start looking for solutions. At the same time the rapid de-carbonisation of the electricity supply will greatly improve the environmental benefits of using electricity as an energy source for cars and other forms of transportation. This is crucial as transport and the use of diesel and petrol is a huge source of carbon emissions and oil consumption.

Peak oil, gas and nuclear?

It is not only environmental problems that mitigate against the indefinite use of fossil fuels. There is also evidence that they may not all continue to be as abundant and easily extractable as many think. The situation is most apparent with oil.

Peak oil is a concept that has entered the public psychology, and put simply it refers to the time at which oil production peaks, after which it will continue to decline. This may be due to technological reasons, for example a move away from oil. It may be because the oil supply cannot keep pace with the growth of demand, or it may be because there is simply not enough oil left to extract. It could also be because we decide we no longer wish to extract it. No one knows exactly when peak oil will happen, but there is evidence that it may be sooner than many previously thought.

It is important to realise that peak oil does not mean that oil will run out in a definitive sense, but simply that the supply cannot be maintained or increased and may start to decline. In a world with projected energy demand increases this means that oil will no longer be able to provide the reliable and affordable source of energy it has until now.

Commentators on peak oil are normally divided into two camps, the optimists and the pessimists. Chief among the optimists has traditionally been the International Energy Agency (IEA) – a Paris-based organisation formed in 1973–74 by the Organisation for Economic Cooperation and Development, with the intent of ensuring a reliable and affordable supply of energy. They have long argued that oil is abundant, both in its conventional form and in its alternative and harder to extract varieties such as tar sands and shale oil or in deep-water oil fields. As cheap oil becomes scarcer, they argue, the price will increase, incentivising the exploitation of unconventional sources. They also point to the continuing discovery of new oil sources, saying that greater investment in exploration will yield new finds, particularly offshore. Recent finds off the Brazilian coast and in Angola are cited as examples of this. In the view of the optimists, the availability of oil is essentially an economic problem, which can be overcome with investment in supply side measures. For some time, this view held sway, and continues to enjoy strong support. In recent years, however, the IEA oil analysis has come under increasing scrutiny, not least because of its unrealistic predictions of ever-increasing demand and supply of oil, with lines on its graphs rising ever upward into the future. In its 2004 World Energy Outlook, the IEA estimated the medium- to long-term price of oil at $22–35 a barrel, with $35/bbl representing the high-price scenario.[20] Both now look hopelessly optimistic (even in the depths of the great recession in 2008, the price of oil never fell below $30/bbl and the US Energy Information Administration estimates average costs for 2011 at $93/bbl).[21] At the same time the IEA has consistently underestimated the growth of the renewable industry, predicting an annual expansion of 4.5 per cent (in fact it has been growing at more like 30 per cent). In 1998 it estimated that global wind energy could reach 47.4 GW by 2020, a figure that was surpassed in 2004.[22] These sorts of predictions matter because the reports from the IEA carry great weight, both with the media and in government circles. It is likely that an excessively cautious view of renewable energy potential has undermined support for emerging technologies, although as we shall see later, this has changed in recent times.

Even as the IEA has grown less bullish on the future of oil, the pessimists have been gaining voice. For their part they think that the peak could be reached much sooner. They point out that new oil discoveries have been steadily declining, and that since the 1980s, the consumption of oil has outstripped new discoveries. Oil fields have a limited lifespan, and after their production peaks, they tend to plateau before declining in output and quality. In order to maintain the supply, new fields or technical means of extraction must be employed. Indeed, many of the main oil fields have already peaked, particularly in non-OPEC countries. US production has been in decline since 1971, the UK since 1999, Norway since 2001. In one of the world's largest oil fields, the Cantarell in Mexico, production has been on the wane since 2000. Fresh discoveries are constantly made, but they cannot be made in sufficient numbers to keep up with an ever-increasing demand. New and more demanding sources of oil must be sought out, deep beneath the sea, or in less accessible forms. Some of these so-called 'alternative' sources of oil have significant risks attached. The tar sands or heavy shale oils in Canada and North America contain huge deposits of hydrocarbons (by some estimates Canada has as much oil as Saudi Arabia). The processes involved in extracting oil from the tar sands are very environmentally destructive, however, involving large-scale mining and heavy energy use needed to 'separate' the oil from the sands. This means that tar sands have a higher carbon dioxide output per barrel than conventional oil and require the destruction of large areas of land. As a result the Canadian government's support for tar sands has drawn condemnation from environmental campaigners and local First Nation's groups. At a 2010 shareholder meeting in London, BP suffered one its largest ever shareholder revolts from institutional investors over the risks associated with its plans to pursue tar sands production.[23]

Deep water oil too has its drawbacks. The recent disaster in the Gulf of Mexico, where a drilling platform exploded, killing eleven workers and causing a massive oil spill, has shown the difficulties of dealing with oil spills in deep water. The spill was eventually brought under control, but not before colossal quantities of oil were released into the ocean. Although the scale of the tragedy is huge, the sad fact is that this disaster has probably occurred in one of the best places it could have. Outside of hurricane season the Gulf of Mexico is a relatively calm stretch of water, with perhaps the best offshore infrastructure in the world. Not only that, but its impact on the USA ensured maximum political pressure on the oil companies involved. Imagine if this spill had happened under the ice of the Arctic, or in international waters of the South Atlantic. It could take months if not years to tackle, and there may be no authority politically capable of ensuring a clean-up. The question for those examining the economics of oil is if the colossal costs of such disasters will be factored into the risks associated with oil production and passed on to the consumer, or if they will ultimately be written off as exceptional events, borne by the tax payer.

As well as the scarcity of conventional oil reserves, peak oil pessimists also cite the lack of transparency on the state of national reserves. Members of the Organization of the Petroleum Exporting Countries (OPEC), a cartel which controls 33 per cent of global oil production and nearly two-thirds of its reserves, have their annual production quotas fixed in part by the size of their proven reserves.[24] This has led to a great deal of secrecy as to the extent of many reserves, and indeed the markets received a shock when a Chief Executive of Saudi Aramco (the world's largest oil producer) was forced to concede that Saudi Arabia's proven reserves were only half as large as had been previously claimed, and that its two largest fields had already passed peak production.

Perhaps the real shift came in 2008, however, when the tone of the IEA began to change and the views of the optimists and the pessimists began to converge. In its 2008 World Energy Outlook, the IEA revised the rate of decline in existing oilfields from 3.7 per cent to 6.7 per cent per year.[25] Then in an interview with British journalist George Monbiot, its Chief Economist admitted that the peak could come sooner than had been previously estimated: 'In terms of non-OPEC, we are expecting that in three or four years' time the production of conventional oil will come to a plateau, and start to decline. In terms of the global picture, assuming that OPEC can invest in a timely manner, global conventional oil can still continue, but we still expect that it will come around in 2020 to a plateau as well ... I think time is not on our side.'[26]

The implications of this seem to have gone largely unnoticed in the wider media, but are deep and wide reaching. Even the IEA is now suggesting that the conventional oil supply could peak in 2020, just eight years away. After this, we will be forced to rely on more expensive and far more environmentally destructive means of oil extraction.

So, whether one is an optimist or a pessimist it seems likely that oil prices will continue to remain high as conventional resources peak. Whether the oil price will be high enough to justify the extraction of unconventional petroleum is unclear, particularly in light of the recent disasters. As the cost of renewable energy falls, and consumers seek alternative energy sources, it may be that many of these oil resources will never be tapped. As has been said many times, the Stone Age did not end for lack of stones. Of course it may equally be that social and political inertia, the ingrained nature of the oil industry and the lobbying power of the companies that benefit from it mitigate against the move away from oil, and society simply pays the price, however high.

From an environmental point of view we simply cannot justify the continue extraction of oil whatever the price, and many 'environmental optimists' see peak oil as a blessing in disguise, a situation which could force powerful vested interests to look again at the alternatives to fossil fuels. While there are other fossil fuels in abundance, the lack of oil will force a radical change in infrastructure regardless of what replaces it, and this will provide an opening for new and more environmentally sound technologies to become established.

While the 'peaking' problem is perhaps most acute in oil, the other non-renewable fuel sources are on the same trajectory, just many (and in some cases many, many) years behind. Of immediate interest is natural gas, or methane. Although the IEA's 2009 World Energy Outlook estimates that proven gas reserves are sufficient to maintain current consumption levels for 60 years, rising demand has led some to believe that conventional gas production could peak and decline by 2025–2040. If natural gas is called upon to replace coal for electricity or oil for transport, demand will sky-rocket and this point could be reached far sooner. The situation is far from clear, however, as gas is relatively unexplored compared with oil. By some estimates there is more than four times as much gas undiscovered than in the current proven reserves, and as with oil there are large quantities of unconventional gas reserves, such as shale gas, which high prices and new techniques are making it possible to recover. Indeed, the recent 'discovery' of this shale gas could be something of a 'game changer', as far as global energy supplies are concerned. The ability to exploit shale gas has caused the USA to overtake Russia in 2010 as the world's largest producer of natural gas, and there are significant potential reserves in many other countries such as China, Argentina, Poland and the UK.

At present it is simply too early to tell what the long-term impact of new shale gas reserves will be. For example there are significant localised environmental concerns that arise from the processes used to extract shale gas ('fracking') and even from a purely 'climate-focused' point of view it is not necessarily good news. While natural gas is significantly cleaner than either coal or oil, producing 60 per cent less carbon dioxide per kWh of electricity than coal, the trouble with 'fracking' is that the process appears to release large quantities of methane – a powerful greenhouse gas in its own right. Some commentators suggest this might actually make 'fracked' natural gas more carbon intensive than coal,[27] although there is still a great deal of debate about that. Even if it is not quite that bad, it seems likely that 'fracked' natural gas has a significantly higher global warming contribution than conventional gas.

Nonetheless, although we cannot afford to burn all the gas that remains, conventional natural gas at least could play an important role as a bridging technology in the transfer to renewables and the speed with which plants can be brought on or offline means it can also provide a useful back-up to renewable energy. For certain applications, such as CSP, it can be coupled directly to the generation process providing small-scale top-ups to ensure the systems run at maximum capacity. On a larger scale, the flexibility of gas, and the speed with which it can be switched on and off, means that it can play an important role in smoothing out the intermittency which some renewables suffer from, until such a time as more advanced grids and storage systems render this unnecessary.

Equally as interesting as peaking fossil fuels is what some are calling 'peak nuclear'. Nuclear energy is often held up as a solution both to peak oil and the climate crisis, since it produces relatively few carbon dioxide emissions and is an established and reliable technology. In order to generate nuclear energy, however, rare elements such as uranium and thorium are required. Standard fission reactors use the isotope U-235 as fuel, which occurs as just 0.72 per cent of all uranium. Other sources of nuclear fuel include thorium (which can be 'converted' into U-233 or used in a thorium reactor) and U-238 (which can be converted into plutonium in fast breeder reactors). All of these elements are scattered widely but thinly throughout the Earth's crust, occurring in useful concentrations in only a few locations.[28]

Of the main nuclear fuel sources, U-235 is by far the most commonly used, accounting for the vast majority of the nuclear energy currently generated. Its supplies are not limitless, however, particularly at economic prices. At current global use, some estimates suggest that the majority of harvestable uranium-235 (up to $130/kg) could be used up by 2035, even with waste reprocessing and the extraction of fuel from nuclear warheads. Even assuming that new discoveries are made, it is estimated that by 2035, around two-thirds of uranium reserves at $130/kg will be exhausted. The International Atomic Energy Agency is more optimistic and estimates that reserves should be sufficient to last 85 years at current rates of consumption. This is still not a great deal, however, as nuclear energy plays an important role in only a few countries (six countries account for over 80 per cent of nuclear energy production), and generates around 13 per cent of global electricity supplies (and falling). Were it to be called upon to meet a far greater proportion of global energy needs it could quickly begin to run out. A tripling of installed nuclear capacity using conventional fission technology could lead to shortages of U-235 fuel within 25 years. Other methods of harvesting U-235 do exist – it can be extracted from sea water for example – but at present these are far more expensive than conventional supplies, costing around $200/kg.[29]

Fast-breeder reactors (which can create their own plutonium fuel from U-238 in the reaction chamber) could extend the available uranium supplies by thousands of years, although they would do so at a higher cost, and are often only considered to be competitive with other nuclear energy sources at uranium prices of over $200/kg. Fast-breeder reactors have so far proved unpopular with investors and despite a few plants in China, Japan and India, at present there is little indication that they are being widely considered for the next generation of nuclear power stations.[30] The fact that they create plutonium as a by-product brings with it obvious security risks, since plutonium is used in the manufacture of nuclear weapons. A massive expansion of nuclear technology, and fast-breeder technology in particular could have serious risks for non-proliferation. Imagine 10,000 new reactors scattered across the globe from Columbia to Korea, Pakistan to South Africa, and the endless shipping lanes transporting fuel and waste between them.

Uranium-based nuclear power is also very expensive, particularly when its associated clean-up costs are included. It has been reported that in the UK alone, the clean-up for its existing fleet of nuclear reactors could be at least £70 billion ($110 billion).[31,32] This is an extraordinary expense for an energy source that supplies just 19 per cent of the UK's electricity. £70 billion put into offshore wind and storage would probably be a far better long-term investment.

The potential bright spot on the nuclear map is the use of thorium-fuelled fission reactors. Thorium technology has been around for a long time but was neglected in favour of uranium-based technology, possibly because of the close links between uranium and nuclear weapons production in the early days. Thorium-based rectors do have a number of important advantages, particularly in terms of safety and non-proliferation, and a large number of advocates. Perhaps the most important feature of a thorium reactor is that it cannot suffer an uncontrolled nuclear reaction or meltdown, nor can it be easily used for nuclear weapons. Nonetheless given the immaturity of this technology it seems likely that in the future renewable energy will be a more cost-effective mechanism of meeting the bulk of global energy requirements in the near term, although thorium certainly seems worthy of future research and investigation. However, while thorium can represent an alternative nuclear fuel source to uranium, it is not inexhaustible, and it too is available only in limited supply. According to some researchers, its position is similar to that of uranium, and supplies can be expected to decline by the middle of this century should it enter widespread use.[33]

Nuclear is a good example of a potential economic peaking of an energy source. It is already amongst the most expensive sources of energy available (particularly when security, insurance and decommissioning are considered). As conventional nuclear fuel becomes scarcer the industry may be forced to turn to even more expensive sources. At the same time the recent nuclear disaster at the Fukushima plant in Japan will demand extra safety measures, and make nuclear even more politically difficult to pursue in the face of understandable public concern (as this book is going to print, Germany has indicated that it will move ahead with closing its nuclear programme while China has indicated it is rethinking its nuclear strategy). It therefore seems unlikely that in the absence of a major technological innovation nuclear will be able to replace fossil fuels as the primary energy source, and instead it seems more likely that it will be called upon to play a supporting role to renewables and natural gas. Similarly, the inability of current nuclear power stations to be ramped up and down to produce power on demand means that they may be unsuitable and expensive for the flexible grids of the twenty-first century in which multiple energy sources will be brought on and offline in response to demand or fluctuating weather conditions.

Energy for development

While we may be growing increasingly concerned about energy security and environmental damage, it is also true that energy is a prerequisite for modern economic development, and for the citizens of poorer countries increased supply will be essential to improving their incomes and opportunities. Certainly, as standards of living improve across much of the world, and particularly in Asia, global energy consumption is increasing rapidly. In 2008, total primary energy consumption worldwide stood at around 474 EJ (~11,295 MTOE),[34] of which electricity consumption accounted for 17.1 per cent (equivalent to around 19,853 TWh). This represents an increase of nearly 50 per cent since 1995, and a doubling since 1971. With the earth's population set to grow to more than 9 billion by 2050 and with continued rapid economic growth in certain countries, this consumption is expected to increase by up to 40 per cent in the next 25 years.

Unfortunately, this rapid growth in energy consumption and prosperity has been accompanied by a huge rise in carbon dioxide emissions. China is now the world's largest emitter of greenhouse gases from energy production, and even on a per capita basis it is broadly equal to some northern European countries like Sweden (Table 1.1 shows a list of the greenhouse gas emissions of selected countries in 2006).

India is already the world's third or fourth largest emitter of greenhouse gases, despite the extreme poverty of much of its population. Other countries, from the Middle East to Ireland (a very rich country, but one which has grown rapidly) have seen rapid rises in emissions as their economies have expanded in recent years. Way out in the lead are a collection of super-polluting countries, whose per capita emissions dwarf those of even other developed states. Chief among these are the USA, Canada, Australia and the Gulf States.

There are some who argue that it is the sole responsibility of the developed world to reduce its consumption of fossil fuels, and in the meantime the

Table 1.1 Total and per capita greenhouse gas emissions from energy consumption in 2009*

Country	Total emissions of CO_2-eq (millions of tonnes CO_2-eq)	Per capita emissions of CO_2-eq (tonnes CO_2-eq / per person/year)
China	7710	5.83
USA	5424	17.67
Russia	1572	11.23
India	1602	1.39
Japan	1097	8.64
Germany	766	9.30
Canada	541	16.15
United Kingdom	520	8.35
South Korea	528	10.88
Sweden	51	5.58
Qatar	67	79.82
Iran	527	6.94
Brazil	420	2.11
Mexico	444	3.99
South Africa	450	9.18
France	397	6.30

* Note that this does not include emissions from land-use change, which are significant. By some estimates Indonesia has at times been the world's third largest emitter of GHG, thanks to deforestation. These figures also ignore the idea that importing countries are partly responsible for the emissions of countries that manufacture the products they consume

Source: US Department of Energy, Energy Information Agency Statistics. http://tonto.eia.doe.gov/cfapps/ipdbproject/iedindex3.cfm?tid=90&pid=44&aid=8&cid=regions&syid=2005&eyid=2009&unit=MTCDPP

developing and recently emerged countries should be allowed to pursue whatever path they can to make them rich, if necessary pursuing the same path as the developed world has followed until recently. There are others who suggest that while it is true that the rich world must dramatically reduce its use of natural resources, and fundamentally readdress the way it deals with the natural world, the solution lies not only in one court. The emissions of the poor and middle-income countries are so high that even if the developed countries ceased to exist, there would still need to be a reappraisal of the use of resources. India alone for example has more people than the whole of the USA, Canada, Europe, Australasia and Japan put together. China or Sub-Saharan Africa have as many again. The truth is that we cannot end poverty without a technological and industrial revolution, as well as a change in lifestyle by the richest. A massive expansion of renewable energy will be key to this transformation.

As an aside this raises an exciting prospect. Since current targets suggest that developed economies require reductions in emissions of 80–90 per cent this would see the per capita carbon dioxide emissions of the European Union or the United States from power generation fall to about 1.8 tonnes per annum and 4 tonnes per annum respectively. In the case of Europe, this means that its emissions must fall to broadly the same level as India's are today, and well below China's. At the same time the per capita emissions of these countries are likely to have expanded rapidly, in accordance with their entitlement to grow. This means that at some point in the next twenty to thirty years they are likely to converge and overlap. It may be of course that emerging nations then overtake the old developed ones in terms of per capita emissions, but equally it raises the hope that this will never happen and that the technologies that have allowed the rich world to reduce its footprint will be accessible to all, enabling a general reduction.

So this brings us to the final point. We know that if we want to halt, or slow down runaway climate change, we cannot continue to burn fossil fuels. We also know that affordable fossil fuels or nuclear will not be available indefinitely. At the same time, if we are to alleviate poverty, we must accept that in many countries at least, energy consumption is going to continue to rise. What we need to do is decouple the rise in energy consumption from the rise in pollution. There are two major ways to achieve this. The first is to use energy more efficiently. The second measure is to generate electricity (and ultimately heating and transport fuel) in a way that will not produce carbon dioxide, or run out.

The role of renewable energy

Currently, with the exception of large hydropower, renewable energy is only just beginning to emerge from the shadows as a mainstream energy source. While in many parts of the world biomass (wood, charcoal, and dung) continues to be a primary source of heat and fuel, and accounts for around 10 per cent of global primary energy needs,[35] this consumption is largely unsustainable and brings with it numerous health and environmental problems.[36] (Forest clearance for charcoal and heating wood for example is a major concern in many parts of the world, devastating biodiversity and promoting desertification, while indoor air pollution from burning wood and animal dung in enclosed spaces reportedly leads to respiratory problems contributing to up to 2 million deaths per year.)[37] As far as modern renewable energy for electricity is concerned, it is only in the last few years that several technologies have really begun to make an impact. Wind power is now the fastest growing electricity source in both the USA and European Union, with more new wind peak capacity installed in 2008 and 2009 than any other form of new electricity.[38] In some markets, such as Denmark, Spain or Germany, wind power now makes a significant contribution to national electricity demand. Other countries, like China and India, are also installing wind power, with China emerging as the largest market for onshore wind power in 2009, with 13,000 MW of new wind power installed that year alone.[39]

Solar too is growing fast, albeit from a much lower base, with the photovoltaic sector expanding by an average of 35 per cent from 2004 to 2008. Despite this, the total contribution of renewable energy to the electricity supply is still small, and there is a long way to go. China may be a leading wind power market, installing over 18 GW in 2010, but in 2009 it also installed around 60 GW of new coal-fired capacity.[40]

There is a window of opportunity, however. In the coming years many of the fossil-fuel powered stations in Europe and North America will reach the end of their lives. If these can be replaced with renewable and low-carbon energy sources, it could make a dramatic contribution towards a more environmentally friendly economy. At the same time, there is a chance for many developing countries to 'get in on the ground floor' with renewable energy, leap-frogging more developed economies and developing a grid and energy source more suitable to the twenty-first century. This 'leap-frogging' phenomenon is already apparent in other key sectors, such as telecoms. Countries which for years struggled with poor telecommunications and the expense of installed fixed lines, are now able to bypass the technology entirely, going straight for wireless and cellular communications.[41] So rapid has this expansion been that even in remote villages, there is likely to be at least one person, if not more, with access to a mobile phone.

Renewable energy technology is also progressing fast and the options for replacing fossil fuel consumption are coming more sharply into focus, with a practical vision of large-scale renewable and low-carbon energy beginning to emerge. Decentralised energy, efficiency and community generation will play an important role, while in the centralised market people are talking grandly of huge super-grids, piping clean energy into developed and developing countries alike. In Europe, vast banks of offshore wind turbines and marine energy systems could stretch from the coast of Portugal up to Norway and Britain and East to the Baltic. In the Mediterranean huge fields of photovoltaic and concentrating solar power generators can tap into the endless sunlight the region is famous for. The much discussed DESERTEC Initiative and Mediterranean Solar Plans have visions of lines of solar installations stretching through the deserts of North Africa, providing power for the emerging economies of the region, and trading electricity with Europe. Elsewhere in the world, from Rajasthan in India to the Gobi desert of China, similar ideas are being touted. Meanwhile in the South West of the United States, in the deserts of New Mexico and California, construction is under way, the first tentative agreements for large-scale utility solar have already been signed, and the race is on to speed up production and deployment.

Solar energy

Of all the available renewable energy sources, it is solar which offers the greatest long-term potential. The Earth receives more energy from the sun in a day than human society uses in nearly a decade. Other renewable energy sources will play an important role, but only solar has the ability to meet and exceed all of our energy needs, even assuming we expand into the future. Estimates based on global energy demand in 2002 suggest that just 4 per cent of desert areas could supply global primary energy needs (some five times more than electricity demand).[42]

Solar power is naturally at its best in sunny places. While some technologies will work in cloudy conditions, their output is significantly less than in full sun. Other technologies will only function in direct sunlight. Fortunately, for much of the world this is not a problem and bright sun can be virtually guaranteed for much of the year. This is particularly true of the so-called 'sunbelt' countries that lie between 30 degrees north and south of the equator. Most suitable of all are desert areas. Almost by definition deserts receive a great deal of sunshine. They are also vast, covering around a third of the land surface on earth. Finally they are sparsely populated, making them potentially ideal for large-scale solar development (albeit with possible downsides for wildlife and habitats).

As has already been mentioned, the basic technologies for producing electricity from the sun have been around for decades. In the case of photovoltaics (PV), this has been since the 1950s, while for concentrating solar thermal power (CSP) the late 1970s and 1980s. Now these two main technologies are expanding rapidly. PV in particular has enjoyed growth of installations of nearly 35 per cent a year for much of the last decade, so by 2009, the annual rate of installation was 7200 MW with a total installed capacity of 23,000 MW (see Chapter 2). CSP started later, but is now about to enter commercial deployment. Total installed capacity at the start of 2010 was only 500–600 MW, but with 2000 MW under construction in Spain alone this will not be the case for long. The real question is not whether solar will play a greater role in the future, the issue is just how much greater. This book will explore both the technologies that can make this happen, but also the projects that are under construction and those that exist as grand plans. There is no sure way of knowing which direction society will choose in the coming years, but large-scale energy from the deserts offers an essential piece of the answer, and one that may make all the difference.

Conclusion

As we begin the second decade of this new century, the problems and challenges facing us are daunting, and it would be foolish to suggest that there will be any easy answers. Yet there is hope. By rediscovering the importance of conserving resources and valuing nature in its own right we can begin to make some of the behavioural changes necessary to overcome our problems. At the same new technology, new means of production and consumption can help to ensure that what we do produce does not fundamentally undermine our future and well-being and our ability to expand prosperity. The advancement of clean, carbon-neutral renewable energy will be a crucial part of that process, and solar will be right at its heart.

Chapter 2

Technology

For the beginner, the world of solar power can be a complex and confusing one. There are a large number of competing technologies designed to harness the energy of the sun and convert it into usable electricity, each complete with its own acronyms, specialist conferences and publications. Each too comes with its own legion of fans and advocates, all prepared to explain that theirs is the technology that really will change the future. Similar sounding names like CSP, CPV, solar thermal and passive solar all add to the confusion, while the mainstream media frequently gloss over the details entirely, making it hard to understand what is going on, or to differentiate between projects announced. Fortunately, the general principles of these technologies are not as complicated as they might sound, and what follows is a basic introduction into the history and technology of solar power for electricity.

Perhaps the first thing to get out of the way is whether the solar power is deployed for producing electricity or heat. While people often use the term to refer solely to electricity producing technologies, strictly speaking solar power can refer to any mechanism for extracting useful energy from the sun. For the purposes of this book, solar technologies will be largely divided into two categories, photovoltaic technologies that 'use' the photons in sunlight to produce a charge, and those solar thermal technologies which focus sunlight to produce heat, which can be used to drive turbines and produce electricity.

It is the second form of solar technology – solar thermal – that tends to cause the most confusion, since it can be used for generating heat, electricity and even to drive air-conditioning, depending on the temperatures generated and the technologies employed. Helpfully the US Energy Information Administration (EIA) divides solar thermal into three categories – low, medium and high temperature.[1] Low temperature systems are typically used for heating swimming pools or similar applications, medium temperature systems are used to provide domestic hot water, while high temperature systems can provide boiling water or steam for factories, electricity generators and other uses.

In domestic applications, the most common form of solar thermal is for providing hot water for washing and showers. This form of small-scale medium-temperature solar thermal is also sometimes known as passive solar. At its simplest it can be a fairly low-tech application, not far removed from a bucket of water left out in the sun – like a bush shower in Africa or Australia. Modern domestic systems, however, are a little more sophisticated, generally involving a series of tubes or coils arranged in panels and mounted on the roof of a building. These 'collect' light from the sun and convert it into heat which is run into a hot tank by means of heat exchanger, where it heats the mains water stored there. Despite the simple principle not all small-scale thermal systems are as basic as they sound, with the more advanced models favoured in rich countries using evacuated tubes (tubes with the air removed to create a vacuum – like a thermos flask) as a highly efficient means of insulation. These systems, which superficially resemble radiators mounted on rooftops, have become a common sight in many countries, and even in gloomy London where I sit, the sharp-eyed observer can spot the rows of thin glass tubes as they travel about the city.

Despite its obvious uses, for the purposes of this book solar thermal for low- and medium-temperature heat will not be considered in any detail, and this chapter will focus on solar technologies for electricity generation using the two categories mentioned above – those that convert sunlight into electricity using semiconductors (photovoltaics or PV), and those that use sunlight for heat, collected via mirrors or dishes to drive a turbine (the high-temperature solar thermal system) – known as concentrating solar power (CSP).

Within these two fields (PV and CSP) there is a further round of subdivisions and many competing technologies, some of which will be outlined below. One of the major practical differences between the two is that CSP requires direct sunlight to work, whereas photovoltaics are generally able to continue producing some power in cloudy and hazy conditions (although they will produce a lot less). Perhaps a more important practical difference is that CSP generators can much more easily store energy in the form of heat, allowing them to continue operating in the absence of direct sunlight. Indeed, some are able to store enough heat to allow them to continue generating electricity for up to 24 hours. Coupled with the ability to use natural gas as a back-up, this makes CSP a potentially predictable and 'bankable' source of renewable electricity.

PV for its part has many advantages too. The lack of moving parts and the durability of PV panels reduces the risk of mechanical breakdown, while the fact that they do not require water for cooling or steam generation could allow them to operate more easily in remote or arid areas. Indeed, aside from regular washing to remove desert dust, they require no water at all (and in fact dry methods of cleaning have also been developed). This may turn out to be a crucial feature in future applications, as water use is emerging as an important issue in the design of large-scale solar energy systems. Finally, while PV remains an expensive technology, prices are falling extremely rapidly, and there is a real hope that in the future new types of solar cells, coupled with economies of scale, will be able to dramatically reduce costs. If this happens then PV could become a dominant technology, particularly in urban and semi-urban environments, where its flexibility will allow it to proliferate into any number of niches and applications.

Discussion of the finer points of solar systems could fill many volumes in its own right, and this book will not attempt to give a detailed understanding of the engineering or physical aspects of these technologies, but will instead seek to provide an overview that will be useful for non-specialist observers.

Concentrating solar power

Concentrating solar power (CSP) is the name given to those technologies that concentrate sunlight to produce heat for electricity. While its use in electricity production is only around 25 years old, the principle of using dishes or mirrors to harness the sun's energy is hardly a new idea. Any child knows that they can set fire to leaves with a magnifying glass, and the idea must be as old as the lens itself. In fact one of the earliest recorded uses of large-scale concentrating solar technology may have been as part of an Ancient Greek defence programme. Writing in the second century BC, the Greek historian Lucian describes how Archimedes used bronze mirrors up to 60 feet across to focus the sun's rays onto ships of the Roman fleet during the siege of Syracuse, causing them to catch fire. While there appears to be significant debate among historians over the accuracy of these comments (it may simply be Ancient Greek propaganda), and it is certainly beyond my expertise, reports suggest that experiments using mirrors similar to those available to the Ancient Greeks have shown them capable of igniting reconstructed wooden vessels.[2] The key, as with modern concentrating technologies, is to focus the sunlight onto a small area, causing it to heat to a very high temperature.[3]

Moving on a couple of thousand years the general idea of the technology remains the same, except that instead of focusing the sun's energy onto an enemy warship, the radiation from the sun is focused on a central receiver containing a heat transfer medium (usually a salt or mineral oil, although experiments are being conducted with steam). This hot medium is then passed through a heat exchanger where the heat is extracted and used to generate steam and drive a turbine. In this way concentrating solar thermal power has been employed to generate large-scale electricity in California since the 1980s, with a total of 11 plants of various kinds constructed between 1985 and 1990. In addition to these functioning plants, the 1980s and 1990s saw the construction of a number of experimental generators in both Europe and the United States, with researchers at Sandia National Laboratory in New Mexico and the Plataforma Solara de Almeria (PSA) in Spain leading the way.[4] Israel too has played an important role in the development of CSP, with early important contributions in the development of Rankine Turbine engines (see paragraph on generators below) and special materials which allowed high temperatures to be achieved in CSP systems. More recently it has played host to several important test sites helping to refine the technology.

In the last few years, these pioneering installations have spawned a new generation of small- to medium-scale commercial CSP plants, which are in turn paving the way for the truly large-scale projects on the horizon. At the same time research is continuous and ongoing, with a number of cost-reduction pathways being explored. This is essential as the fast changing nature of the global renewable sector has put many businesses under pressure, particularly in light of falling PV costs and the global financial crisis.

> *Box 2.1* **A note on names**
>
> The names of solar systems tend to divide into three general categories. There are those with inspirational names, like Sunrise, those simply named after places, like Andasol I and Andasol II, and those that just have letters and number, such as PS10. Taken together this leads to a swirl of unhelpful, similar sounding and often confusing names for projects. There is no real way around this. I have included tables of most of the existing and planned systems in Chapter 5, but in the meantime, try not to get put off by the many titles in the passages below, they are just the names of solar plants included to illustrate the development of the sector.

Within the broad category of concentrating solar power there are a number of varying technologies. These are often categorised as parabolic troughs, solar towers, dish–Stirling systems and linear Fresnel reflectors – depending on the mechanism for collecting solar energy or the general mode of its use. There are also a number of additional components, such as storage systems and natural gas back-up which are crucial to the economic viability of CSP, and these are outlined below. Finally there are solar chimneys, which are not always considered CSP and are distinguished by their lack of mirrors or a thermal cycle, but nevertheless are listed for completeness, since they frequently appear in this context. Table 2.1 summarises the performance of various CSP technologies.

Thermal cycle / generator

The thermal cycle, or generator, is the heart of any CSP system. This is the bit that actually does the work and drives the turbine. The purpose of the

Table 2.1 Reported and estimated performance of various CSP technologies[a,b]

	Capacity per unit (MW)	Concentration of sun	Peak solar efficiency	Annual solar efficiency	Thermal cycle efficiency	Capacity factor (solar)	Land use m²/ MWh/yr	Levelised costs (€/kWh)[c] estimated or actual 2005
Parabolic trough	10–200	×70–80	21% (d)	10–15% (d) 17–18% (p)	30–40% ST	24% (d) 25–90% (p)	6–8	0.17–0.19
Linear Fresnel	10–200	×25–100	20% (p)	9–11% (p)	30–40% ST	25–90% (p)	4–6	0.16
Power tower	10–150	×300–1000	20% (d) 35% (p)	8–10% (d) 15–25% (p)	30–40% ST 45–55% CC	25–90% (p)	8–12	0.13–0.18
Dish–Stirling	0.01–0.04	×1000–3000	29% (d)	16–18% (d) 18–23% (p)	30–40% Stir 20–30% GT	25% (p)	8–12	0.38

(d) = demonstrated, (p) = projected, ST = steam turbine, GT = gas turbine, Stir = Stirling Engine, CC = combined cycle, solar efficiency = net power generation/ incident beam radiation, capacity factor = solar operating hours per year/8760 hours per year

a MED-CSP Concentrating Solar Power for the Mediterranean Region report by DLR – Final Report by German Aerospace Centre (DLR), Institute of Technical Thermodynamics Section Systems Analysis and Technology Assessment.
b Robert Pitz-Paal, ECOSTAR – European Concentrated Solar Thermal Road Mapping; http://www.vgb.org/data/vgborb_/Forschung/roadmap252.pdf
c Robert Pitz-Paal, ECOSTAR – European Concentrated Solar Thermal Road Mapping; http://www.vgb.org/data/vgborb_/Forschung/roadmap252.pdf

mirrors, storage systems and collectors is to supply heat to the thermal cycle. For this reason, the thermal cycles of CSP systems tend to be much the same as those found in other forms of thermal electricity generation, such as coal, gas or nuclear. A brief outline of these systems will help with understanding the storage and cooling technologies that are vital to the economics of CSP installations.

The most common form of thermal cycle in CSP installations is a Rankine cycle, a standard steam–water cycle used in power generation all over the world. This is a closed system, where the water boiled into steam is contained in a loop and is recycled. A basic system has four elements – a pump, a boiler (where the steam is generated), a turbine and a condenser. Water is pressurised by the pump before being heated to steam in the boiler (this is where the solar heat comes in, replacing or augmenting fossil fuels). This steam is then expanded in the turbine, generating electricity, while the depressurised steam is cooled in the condenser before being recycled back to the beginning.

In contrast to the 'wet' Rankine cycle, the Brayton cycle uses compressed air to drive a turbine. The principle is similar – the air is compressed before being heated and expanded with the hot gas used to drive a turbine. The hot exhaust gas is then expelled.

A combined cycle system integrates these two kinds of cycles. Here a Brayton cycle provides the initial generation, while a second Rankine cycle makes use of the hot exhaust gas to drive a steam turbine. Combined cycle systems are common in modern gas power stations and many of the CSP plants which are being developed use them. In this case the heat from the solar is generally used to augment that supplied by the exhaust from the Brayton system in the Rankine cycle. This is known as an Integrated Solar and Combined Cycle System (ISCCS).

Finally there are Stirling engines. First invented in 1815 by Robert Stirling as an alternative to the steam engine, Stirling engines are known as 'external combustion' engines, because the heat source is external to the engine. Like a standard car engine, they work by using the expansion of gas as it heats to drive a piston. While in a car this hot gas is provided by burning fossil fuels, in the case of a Stirling system the gas is already in the engine, in a sealed container. This sealed system is divided into two parts – a cooled contracting part and a heated expanding part. When heat is applied (from the concentrated solar radiation for example) the continuous expansion and contraction of the gas can be used to drive a turbine and hence to generate electricity.

Modern Stirling engines usually use hydrogen or helium as the sealed gas, and are efficient, achieving energy conversion rates of around 25 per cent, although in some cases it can be up to 40 per cent. In recent years they have been coming back into fashion for a number of applications, including micro-Combined Heat and Power as well as Concentrating Solar Power.

Storage technology

One of the crucial advantages of CSP over some other forms of renewable energy, including photovoltaics, is the relative ease of energy storage. As a rule electricity is hard to store, and requires expensive external systems like batteries, flywheels or pumped storage facilities. By comparison, heat is easy to store. Everyone knows that a well-insulated hot water tank or flask will stay warm for hours, and CSP installations are no exception. By storing their energy as heat, some CSP plants can continue to operate in the event of an unexpected change in the weather or mechanical failure. In fact some can store so much heat that they can theoretically continue producing electricity right throughout the night.

While 24-hour generation is possible in the right conditions, in practice many developers are opting for more limited storage systems, using a range of mechanisms – each with its own advantages and disadvantages. Some designs favour molten salt systems, which achieve high temperatures and allow the plant to operate for hours without direct sunlight. A working example of this type of storage is the 50 MW Andasol I parabolic trough plant in Spain. This system uses two salt storage tanks – one at a relatively cool 290 °C, the other at 390 °C. The mineral oil that is heated by the sun is passed through a heat exchanger with the cool salts. These are heated up and transferred to the hot tank until they are needed. To extract the heat, the hot salts are passed back through the heat exchanger to heat the mineral oil, which can then produce steam for the generator. The now 'cool' salts are then returned to the cool tank, where they are ready to be reheated.[5]

Other thermal storage designs include keeping hot water in thermally clad tanks that can then be re-heated by gas or solar energy. This allows the system to run at partial load (without sunlight or with reduced sunlight) for half an hour or so, depending on the volume of the storage tank, long enough in most cases to avoid penalties from the grid operators in the event of an unexpected change in weather. While overall availability may not be as high with a water storage system as with a salt storage set-up, supporters of this more limited storage point out that high-temperature systems suffer more stress than lower-temperature generators, and so are more likely to suffer mechanical failure, potentially increasing downtime and maintenance costs.

Another system, employed by the PS10 power tower in Spain, stores excess steam directly in pressurised tanks, allowing it to run in the absence of sunlight. Steam storage is expensive, however, due to the temperatures and pressures involved, so storage volumes are normally limited to less than an hour without sunlight.[6] Despite this, the technology is considered to be reliable and mature, and so 'low-risk' from the point of view of a technical malfunction. Opting for low-risk technologies is an important compromise which developers in many areas usually need to take in order to secure investment.

Air storage tanks are another option for CSP heat storage. Several prototype devices have experimented with using ceramics or concrete as a heat 'store'. In these systems hot air or liquid from the CSP station is pumped through a series of pipes in a porous storage medium which heats up to around 400–500°C. When the heat is needed again, air can be piped back through the storage block, recovering the energy. The advantage of this system is its relative simplicity. Storage time, however, is currently limited to around 3–6 hours.[7]

Finally, a number of new storage possibilities are under development, including using phase-changing mediums to exploit the heating and cooling point of various salts such as sodium and potassium nitrate. While these systems are still experimental it has been suggested that the low cost of the materials and relative ease of storage could make them economical in the future.

Whatever the systems that are employed, the ability to operate for several hours in the absence of direct sunlight could be of absolutely crucial importance to the large-scale deployment of Concentrating Solar Power, as it enables it to overcome the intermittency issues that affect some other renewable electricity sources, and provide smooth electricity outputs. It is also important commercially as producers that are able to provide peak power can command a premium price, while those that fail to generate the electricity they have promised to the grid can face hefty fines. Thus the ability to generate in the event of an unexpected change in weather or some other failure has wide implications for the usability and cost of electricity generated by CSP. In fact it is this 'predictability' as much as any other factor that could make CSP a viable prospect in the sunbelt countries of the world.

Back-up

Another crucial feature of concentrating solar power is its ability to use fossil fuels such as coal, biomass or natural gas as a back-up. This can greatly improve the availability of the plant. Gas is the most common fossil fuel in use and can either be used to 'prime' the system, or as a top-up. Looking at the production profile of the first parabolic trough systems in California, it is easy to see the importance of this gas back-up. In most years the system used an average of about 8 per cent of its electricity to come from natural gas to ensure that it was operating at maximum capacity. In one or two years, however, the production from the solar component fell. In these situations the back-up was able to step up production, covering around 20 per cent of total generation and enabling it to keep operating at full output. For this reason, many of the plants currently under development have a natural gas component. In these plants the natural gas contribution is fairly small, around 10–25 per cent of the total heat production, and many governments have placed strict caps on the amount of gas that can be burned before the plant can no longer be considered 'renewable', and so be eligible for feed-in tariffs or other support policies.

In some cases, the solar component is more of an auxiliary system which feeds solar heat directly into larger, combined-cycle gas powered generators. These are known as Integrated Solar and Combined Cycle Systems (ISCCS – see above), a type of cogeneration plant. In this situation the fossil fuels are used in a standard internal combustion turbine (gas for example is burnt in a gas turbine), and the waste heat is used to drive a steam turbine. In an ISCCS system, the heat from the solar is fed into the steam turbine component. In these systems the solar thermal element is usually the smaller partner in generation, perhaps 20 MWe in a 200 MWe system. Examples of this type of installation can be found at the new plants under construction in Morocco, Algeria and Egypt, as well as Florida in the USA.

Cooling

All thermal power stations that use steam powered generators require some form of cooling and concentrating solar thermal is no exception (with the exception of dish–Stirling and solar chimney systems). In a standard Rankine steam cycle the steam is maintained in a closed loop system and is repeatedly heated and cooled. Heat is input at a high temperature (source) and put out at a low temperature (sink temperature). The difference between the two represents the work done by the turbine and the theoretical maximum efficiency is determined by the ratio between them. Lowering the sink temperature or increasing the source temperature will generally increase the efficiency of the system. In many cases this cooling requires large amounts of water, but as pressure on water supplies increases, new solutions are being found. These various types of cooling system are generally referred to as once-through cooling, recirculative cooling, wet–dry cooling or dry cooling, depending on the amount of water required. The main forms of cooling systems are outlined below.

Once-through water cooling

Once-through water cooling basically means extracting water from a source, such as a river, using it to cool the generator and then returning it to the water source. While it does not 'consume' any water as such it does raise the temperature of the water in the source. This can have serious negative consequences for aquatic life, which tends to have a narrow tolerance for temperature variations. In addition warmer water holds less oxygen than colder water and has different chemical interactions, both things which can seriously impact on aquatic ecology. Living creatures may also be killed or injured by being pulled through the cooling mechanism.[8] Because of this once-through cooling systems have become highly regulated and alternatives are being sought. This form of cooling is of limited interest to CSP in any case, as there are generally few large water sources like this in the desert – aside from some large rivers like the Nile or Colorado.

Table 2.2 Water use of CSP technologies, including washing and cleaning

Technology	Cooling	Gallons / MWh	Performance penalty	Cost penalty
Coal/Nuclear	1. Once-through 2. Recirculating 3. Air-cooling	23,000–27,000 400–750 50–65		
Natural gas	Recirculating	200		
Power tower	1. Recirculating 2. Hybrid 3. Air cooling	500–750 90–250 90	1–3% 1.3%	5%
Parabolic trough	1. Recirculating 2. Hybrid 3. Air cooling	800 100–450 78	1–4% 4.5–5%	8% 2–9%
Dish–Stirling engine	Mirror washing only	20		
Linear Fresnel reflector	Recirculating	1000 (estimated)		

Source: US Department of Energy, *Concentrating Solar Power Commercial Application Study: Reducing Water Consumption of Concentrating Solar Power Electricity Generation*, Report to Congress.

Evaporative water cooling / recirculative cooling

This is probably the type of cooling most commonly associated with power generation, involving the large concrete cooling towers many people will be familiar with. In these towers, waste heat from the generator is dissipated by evaporating water into the air through the tower. As the water rises and cools in the towers, much of it condenses and runs back into the system, although a good deal is also released as steam. This uses a significant quantity of water (see Table 2.2) and has a couple of other consequences. Firstly, as the water evaporates it may leave impurities behind in the remaining water (these may be dissolved chemicals from the water itself, such as salts or pollutants). These build up over time requiring some of the water to be filtered to remove them. Secondly, as the water evaporates it may take with it tiny particles less than 10 microns across (PM10s). These are a recognised health hazard and are regulated by law in most countries.[9] Nonetheless it is a standard technology and several of the parabolic troughs in use today employ evaporative cooling.

Dry cooling or air cooling

In dry-cooling systems the waste heat from the power plant is ejected into the surrounding air through a heat exchanger. In these exchangers the waste 'heat' is passed through an array of tubes over which air is blown by a fan. Since a significant temperature difference is required for the heat exchanger to work well, the temperature of the exchanger must normally be at least 16–27 °C (30–50 °F) above the surrounding air temperature.[10] On hot days this can result in a higher condensate temperature (waste steam from the turbine) and this can reduce the efficiency of the generator. As a result dry-cooling systems are generally more expensive and less efficient than wet-cooling systems (although to some extent this must depend on the cost of the water). The fact that the efficiency of a dry-cooling system decreases with higher ambient air temperature is of significance for CSP applications as their desert locations are often very hot, and their power is often most in demand to meet peak loads caused by air-conditioning on hot days. A study compiled by the US Department of Energy indicated that output from a parabolic trough in California using dry cooling fell by 5 per cent over the year compared with a wet-cooling system, raising the cost of the electricity by a significant 7–9 per cent. Interestingly, the electricity produced by systems trialled in New Mexico was expected to be only 2 per cent more expensive, since the ambient air temperature at the test sites was significantly lower than in California.[11]

Similarly, the hotter the input temperature of the system, the lower the losses in efficiency and hence the lower the electricity costs. A power tower running at 600 °C for example would be expected to suffer significantly lower efficiency losses than a parabolic trough at 390 °C since the differential with the air temperature will be higher. This could in turn have consequences for the types of systems adopted and where they are deployed.

Hybrid wet–dry cooling systems

Hybrid cooling systems for CSP applications tend to involve a two-stage solution, mixing air cooling with evaporative cooling. In these systems a dry-cooling system involving a heat exchanger (like the one described above) will be the main source of cooling. On hot days, however, a proportion of the condensate will be diverted to a cooling tower, relieving pressure on the air-cooling system and allowing the turbine to maintain optimum performance. Hybrid systems will generally have a small cooling tower and a smaller than normal air-cooling system. While they will still be more expensive than a wet-cooling system the US Department of Energy estimates that they should still be cheaper than relying on an air-cooled system, particularly in hot locations. Hybrid systems can operate at only 10 per cent of the water use of a traditional wet-cooling system.[12]

Parabolic troughs

The most widely deployed of the current CSP technologies, and the one which looks set to take the largest share of the market in the next few years, is the parabolic trough system. Essentially, these consist of long rows

Figure 2.1 Parabolic trough plant in the US (source: Warren Gretz/NREL).

of concave mirrors mounted on tracking systems and arranged into blocks (Figure 2.1). Each line of mirrors focuses the sun's rays onto a central pipe or tube containing a heat transfer fluid, normally mineral oil. This fluid is then heated to a high temperature (generally 300–400 °C) before being passed through a heat exchanger. In this way steam can be generated, which is used to power a turbine (usually a Rankine-cycle turbine) in very much the same way as a traditional thermal power station – the heat generated from burning coal or gas is simply replaced with heat derived from the sun.

At present the largest working example of the parabolic trough is the famous 344 MW array at Kramer Junction in California, developed by the US–Israeli firm LUZ Engineering between 1984 and 1992. Altogether, LUZ built nine plants at this site, ranging from 14 MWe to 80 MWe until the company ceased trading in the early 1990s. The troughs, called Solar Energy Generating Systems (SEGS I–IX, see Table 2.3) continue to function well, and produce electricity reliably. A recent report by the group SolarPaces estimated that their availability exceeded 99 per cent, and that their output had declined by only around 3 per cent since they were installed. Throughout their 20-year history, these plants have produced almost 9 TWh of electricity, around half of all the electricity generated by solar power worldwide. Even today they continue to provide around 800 GWh of electricity per year, enough to power around 160,000 average homes.

Trough collector systems

One of the most obvious and visible parts of a parabolic trough system is its collectors. These consist of vast rows of frames that hold the highly polished mirrors and track the sun's movement from east to west throughout the

Table 2.3 SEGS I–IX history and operational statistics (all in California)

Plant	Year built	Location	Turbine capacity	Collector field area (m²)	Oil temperature	Average annual solar electricity production 1998–2002 (MWh)
SEGS I	1984	Daggett	14 MW	82,960	307 °C	16,500
SEGS II	1985	Daggett	30 MW	165,376	316 °C	32,500
SEGS III	1986	Kramer Jct.	30 MW	230,300	349 °C	68,555
SEGS IV	1986	Kramer Jct.	30 MW	230,300	349 °C	68,278
SEGS V	1987	Kramer Jct.	30 MW	233,120	349 °C	72,879
SEGS VI	1988	Kramer Jct.	30 MW	188,000	349 °C	67,758
SEGS VII	1988	Kramer Jct.	30 MW	194,280	391 °C	65,048
SEGS VIII	1989	Harper Lake	80 MW	464,340	391 °C	137,990
SEGS IX	1990	Harper Lake	80 MW	483,960	391 °C	125,036

Source: National Renewable Energy Laboratory Solarpaces Resource – http://www.nrel.gov/csp/solarpaces; http://www.californiaphoton.com/entities/private/plants/SEGS.html

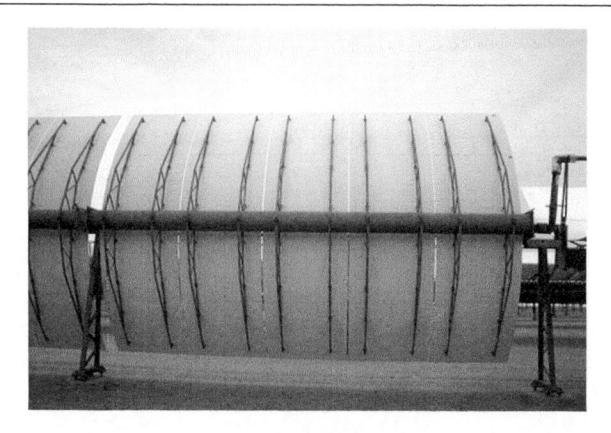

Figure 2.2 LS-2 collector with torque tube structure (source: Henry Price/NREL).

day and focus its energy onto a tube that runs the length of the array. These systems need to be tough, both to handle the constant high temperatures and movement of hot fluids, and also to withstand the desert elements with high winds and sand. Several main designs of system are currently deployed, reflecting the organisations that developed them. Of these the Luz system, the EuroTrough, Solargenix system and Direct Steam System are mentioned below, since they help to illustrate various aspects of the development of the technology. Other designs are certain to come to the fore in coming years, as more and more companies enter the market, each with its own subtly different take on the parabolic trough collector.

LUZ SYSTEM COLLECTORS

As the earliest and longest running of the parabolic troughs, the system collectors designed by Luz Engineering in the 1980s and 1990s represent the standard by which all other collectors are compared. The robust nature of these collectors – made from galvanised steel – makes them tough, and they have proved to be highly reliable. For example, most of the SEGS at Kramer Junction that have reportedly been in daily operation since the 1980s use Luz system collectors, and these continue to produce electricity with a high degree of availability.

There are two types of Luz system collectors in use – the LS-2 and LS-3. The LS-2 collector (Figure 2.2) features a complex design, requiring great accuracy of manufacturing. Its torque-tube structure in the tracking systems (like those found in four-wheel drive cars) provides a good measure of structural stiffness. It has six collector modules, three on either side of the drive.[13]

Crucially, the torque tube uses a lot of steel, which is expensive, and requires precise manufacturing to build. To try and combat this and help

Figure 2.3 Truss support structure from LS-3 parabolic trough collector (source: Henry Price/NREL).

reduce manufacturing costs, Luz designed the larger LS-3 (Figure 2.3) to lower manufacturing tolerance and use less steel. It proved to be a very reliable design. The LS-3 uses a bridge truss structure in place of the torque-tube. Luz's LS-3 collector has truss assemblies on either side of the drive. Each LS-3 truss assembly has three 4-metre-long receivers.[14]

Interestingly, the successor company to Luz Engineering, Brightsource Energy, has chosen not to develop parabolic trough technology at present, and is instead investing in power tower technology. To what extent this decision has been influenced by the experience of designing and manufacturing parabolic trough collectors is unclear, but the company obviously thinks that the commercial opportunities are greater with a power tower design (see below).

EUROTROUGH COLLECTOR

Following the departure of Luz, a European consortium – EuroTrough – initiated the development of a new collector design intended to build on the advantages of the LS-2 and the LS-3. The EuroTrough has been designed to be compatible with the SEGS systems in California, and has been tested both at SEGS V in California (see Table 2.3) and at the Plataforma de Solar Almeria test facility in Spain. The consortium behind the design, which is made up of companies like Abengoa, Flabeg Solar International and Slaich Bergmann and Partner,[15] is offering the EuroTrough for new projects, and it has so far been installed at a recently completed 50 MW trough system in Spain and is being used for the 25 MW Kuraymat project in Egypt. In fact, given the importance of Spanish company Abengoa in the current CSP boom, it seems possible that the EuroTrough will quickly become the most common template for parabolic trough systems.

Figure 2.4 The space frame structure used by Solargenix (source: Tim Wendelin/ Solargenix/NREL).

SOLARGENIX COLLECTOR

Under the US Department of Energy's USA Trough Initiative, a company called Solargenix Energy (formerly Duke Energy, and now a subsidiary of Spanish giant Acciona) developed a new collector frame structure through a cost-shared research and development contract with the USA's renowned National Renewable Energy Laboratory.[16]

The Solargenix collector is made from extruded aluminum, as opposed to the galvanised steel used in the Luz and EuroTrough designs. It uses an 'organic' webbed hubbing structure, developed by Californian company Gossamer Space Frames. The manufacturers claim this new design is lighter and more easily assembled, and has a simpler manufacturing process, than the Luz and EuroTrough systems.[17]

The first large CSP project for over ten years, the 64 MWe Nevada Solar One parabolic trough project (not to be confused with the power tower Solar One – see below), features the Solargenix SGX-1 collector (Figure 2.4) and a similar design has been selected for a number of projects that are in the pipeline.[18]

DIRECT STEAM SYSTEMS (DISS)

Alongside the three tested and functioning collector systems there are various experimental designs under investigation. One of these is a direct steam generating parabolic trough (DISS), in which the transfer medium is replaced with steam, which is heated directly – removing the need for a mineral oil transfer medium. This could potentially lead to significant cost reductions over existing designs, with some estimates placing this at up to 26 per cent.[19] To date the technology has been demonstrated on a 700 metre loop at the PSA test facility in Spain, where it ran at 400 °C and 100 bar of pressure. There are currently plans to construct a 5 MW demonstration plant, known as the INDITEP project.[20]

As with all new developments in this field, any new technology is likely to be 5 per cent innovation and 95 per cent compromise between cost, reliability and technology.[21] The key as ever will be which of these ideas can provide electricity reliably and at the lowest cost, and without running into environmental, resource or regulatory constraints.

Mirrors or reflectors

At the heart of any parabolic trough generator (and indeed all other CSP technologies) are the mirrors required to focus the sunlight. Currently the majority of trough power plants use glass mirror panels manufactured by Flabeg Solar International of Germany (also one of the partners of EuroTrough). Those employed in the LS-3 collectors in California are typical, with each mirror panel 2 metres square and with a second-surface silvered glass layer (meaning that the reflective silver layer is on the rear of the glass). The glass is a 4 millimetre thick, low iron or white glass with a high transmittance, giving a solar-reflectivity of about 93.5 per cent. To give some idea of the number of mirrors involved in a functioning parabolic trough plant, the LS-3 collector features 224 mirror panels on each solar collector assembly. The 80 MWe SEGS IX power plant has 888 LS-3 solar collector assemblies adding up to almost 200,000 mirror panels.[22]

The glass mirror panels have reportedly performed very well during the operation of the Californian SEGS (Solar Energy Generating System) power plants. They have maintained high reflectivity and suffer low annual breakage rates. However, mirror breakage does occur, particularly due to wind stress around the edge of the solar fields. In fact mirrors are reported to be one of the most expensive components of both parabolic trough and power tower generators, and it will be important for the industry to develop ways of reducing costs in order to drive down the price of electricity.

Already a number of alternative mirror concepts are being readied for commercial use, using polymers or deposited layers. One option, developed by the US company ReflecTech, uses layers of polymers with an inner layer of reflective silver. These panels are both lighter and stronger than glass. Even accounting for mounting the mirrors on aluminium frames, the company suggest that the mirrors could be around 50 per cent cheaper than standard glass units, although there is some indication that their reflectivity may be around 5 per cent less than the best glass mirrors. Another option, put forward by the German company Alanod Solar, uses metal alloy mirrors coated in a nano-composite layer. Again, these mirrors could be lighter and cheaper than the glass alternative. Extensive testing of alternative mirror concepts such as these is currently under way at both NREL and the German Aerospace Institute (DLR), with weather testing, abrasion and reflectivity all under the spotlight. If it turns out that these products can compete with glass on durability and reflectivity at a lower price they could represent a significant step forward in the economics of CSP.[23]

Another hope, aside from technological change, is that the sheer scale of the emerging CSP market will bring with it economies of scale and standardisation, since at present mirrors must be specially ordered and designed, project by project, raising costs. The power tower technology being developed by eSolar in California uses larger numbers of smaller mirrors that are of a standard size and shape, and are more easily mounted than larger more individualised panels.[24] This is designed to reduce construction, shipping and handling costs, without any particular change to the make-up of the mirror (for more information, see Chapter 5).

Central receivers or power towers

Like parabolic troughs, power towers use fields of mirrors to reflect the sun's light onto a central receiver; however, instead of being a horizontal tube running along a field of mirrors, the receiver is a single unit mounted on top of a tall tower. As in parabolic troughs this receiver contains a medium that heats up to very high temperatures (600–1000 °C), which is then used to generate steam in a turbine. Once again this is usually a standard Rankine-cycle turbine of the kind used in electrical generators all over the world. In the power tower systems so far constructed, the absorptive medium is usually a mixture of sodium and potassium nitrate (saltpetre). In some of the next generation of towers planned, new heat transfer systems using liquid graphite (a form of graphite in suspension) are being employed. One of the advantages of the power tower is that the high temperature produced can lead to a more efficient thermal cycle than that in parabolic troughs or linear Fresnel reflectors. Also, the fact that the heat transfer medium does not need to be piped through long rows of mirrors may reduce mechanical risk in the system.

The world's first large power tower, Solar One, was built in the Mojave Desert near Barstow in California and had a peak capacity of 10 MW. Designed by the consortium including the US government Department of Energy, Southern California Edison, Los Angeles Department of Water and Power, and California Energy Commission, it was completed in 1981 following a competition to find the best mirror or 'heliostat' to focus the sun onto the receiver. A number of different designs were entered, each involving trade-offs between expense, durability, land area required and simplicity of design. In the end a design produced by Flabeg Solar International of Germany was selected and a system consisting of a total of 1818 mirrors, with a surface area of over 75,000 square metres, was used. Throughout its four-year run from 1982 to 1986 Solar One generated electricity, proving the viability of

38 Desert Energy

Figure 2.5 Experimental solar power tower in California (source: DOE/NREL/Sandia National Laboratories).

the technology. In fact the installation reportedly became something of a landmark in the local area, in part because of its exotic appearance and the strange optical effect it had, with beams of light rising up through the desert (Figure 2.5).

In 1995, nine years after Solar One was mothballed, it was converted into a second prototype called Solar Two. At about 10 MWe, this had a similar capacity to Solar One, but with a number of important differences. Firstly, extra mirrors were added to increase the total size of the field to 82,750 m^2. Most importantly, Solar Two experimented with using molten salt as the heat storage system, allowing it to run for 24 hours a day during some of the trials. The salt used was composed of 60 per cent sodium nitrate and 40 per cent potassium nitrate. The deployment of mineral salt as a heat storage system has been adopted by some of the second generation of commercial power towers currently under development in Spain and the USA, as well as some of the parabolic trough systems.[25]

Solar Two operated from 1995 to 1999, after which it was put on ice, before finally being demolished in 2009, reportedly after a buyer could not be found.

Current power tower design and technology

The second generation of power towers that have recently been completed, and those now under construction, closely follow the technology trialled in

California. Completed in 2006, PS10 is located near the town of Sanlucar la Mayor in the Spanish region of Andalucia and is the first power tower to be built since Solar Two in California. Despite its name, PS10 is actually an 11 MW installation, with a field of 624 heliostats spread over several hectares. As mentioned in the section on storage above, PS10 uses pressurised steam as a storage system (285 °C at 50 bar). This enables the system to respond to peak demand and also to keep running for about 30 minutes in the event of a sudden change in circumstance – enough to avoid penalties from the grid operator.

In addition to storage, PS10 also employs a natural gas back-up system, generating 10–15 per cent of its electricity in this way. Although it has only been in operation for a short time, PS10 has so far operated well, generating around 24.3 GWh per year – a capacity factor of 25 per cent.[26]

Just next to PS10 (quite literally) is a 20 MW power tower called PS20 which came online in 2009. Constructed by Abener Energia for Abengoa Solar, PS20 is the largest solar tower yet constructed. Technologically, PS20 is very much just a scaled-up version of PS10, although reportedly with significant improvements in control and operation. Together, these two installations represent the largest concentration of power tower technology yet built. Current plans are to construct a total of 300 MW of tower systems around Sanlucar in the coming years, presumably providing a significant portion of the region's energy needs.

Looking to the future, the Solar Tres project is likely to become the world's first commercial power tower to utilise a molten salt storage system. This 15 MW plant is due to be constructed near the town of Ecija in Spanish Andalusia. Although closely modelled on the Solar Two prototype in the USA, there are a number of important differences. Firstly it uses a greatly increased solar field, nearly three times the size of Solar Two. This should increase availability and over-all plant efficiency by around 6 per cent. Solar Tres will also employ a large storage system, with capacity for up to 6250 tonnes of molten nitrate salt. This should be enough to provide 500 MWh of electricity, or 16 hours of full running capacity. This means that throughout summer the plant should be able to operate for 24 hours a day, leading to an impressive annual capacity factor of 65 per cent, easily on a par with most modern thermal power stations and similar to many of the most efficient large hydro installations (the giant Itapúa dam in Brazil has a capacity factor of 75–77 per cent). If Solar Tres is able to achieve such a high capacity factor it will help to bear out the predictions on capacity factors made by the German Aerospace Institute in Table 2.1, made in 2005, where potential factors of 25 per cent up to a possible 90 per cent were envisaged.

Dish–Stirling systems

The third main type of concentrating solar power system is the dish–Stirling system. Superficially resembling a large satellite dish, these systems consist

Figure 2.6 Dish–Stirling system in Maricopa (source: David Hicks/NREL).

of a mirror, or dish of mirrors, that focuses the sun's energy onto a Stirling engine (see above), a type of external combustion engine (Figure 2.6). Unlike parabolic troughs or power towers, dish–Stirling systems are highly modular, consisting of large numbers of small turbines with capacities measured in kilowatts, rather than megawatts.

Modern development of the dish–Stirling system was initiated in the mid-1970s by a group of US companies including Ford Advanced Aerospace Development Operations, Boeing Aerospace and McDonnell Douglas Aerospace and Defence. In 1996 a US company called Stirling Energy Systems (SES) acquired the patents for the manufacture of dish–Stirling systems and entered into a long-term agreement with Sandia National Laboratory in New Mexico and the US Department of Energy for further research and development. Stirling Energy Systems' first outing into commercialising its technology was a small demonstration array of six 25 kW systems constructed at Sandia, following which in 2008 it unveiled its modular 25 kWp Suncatcher system.[27]

Along with its distributor arm Terressa Solar, Stirling Energy Systems has recently finished work on a 1.5 MW demonstration power plant near Maricopa in Arizona. Completed in January 2010, the plant employs 60 Suncatcher dishes, and is the first megawatt-scale dish–Stirling array ever constructed. Looking to the long term, Stirling Energy Systems has signed agreements to provide over 750 MW of electricity to San Diego Gas & Electric, a Californian utility (see Chapter 5). Now called the Imperial Valley project, this project will be split into two parts, the first consisting of 300 MW, the second of a further 450 MW. Work on the first phase was originally scheduled to begin in 2010, although recent speculation that the technology may be changed from Stirling Technology to photovoltaic may delay this, while the second phase will be dependent on the completion of the Sunrise Powerlink, a grid extension planned by the California Public Utilities Commission.

Alongside Imperial Valley, SES has filed an application for Calico Solar (formerly Solar One), another 850 MW installation in the Mojave Desert near Barstow in California. Covering 8200 acres, Stirling claim Calico Solar would be able to supply enough electricity to power around 600,000 US homes, much of it at peak times. In anticipation of this project SES has signed a power purchase agreement with another large utility, Southern California Edison.

If they are ever completed, Calico Solar and Imperial Valley will represent a dramatic breakthrough for dish–Stirling technology and for CSP in general. The combined installed capacity of 1.6 GW would be a major achievement and an example of what CSP is capable of.

Three important practical features distinguish dish–Stirling systems from other forms of CSP technology. The first is their inability to use heat storage in the same way as parabolic troughs, power towers or linear Fresnel reflectors. This could have important implications for the areas of the world where they will be suitable, since sunlight is more important than ever. Similarly, they do not use natural gas as a back-up, potentially making them more vulnerable to variability. On the positive side, other forms of CSP require substantial quantities of water to cool the condensers in their generators (as do conventional fossil-fuel power stations). In contrast, dish–Stirling systems do not have external generators and so their only need for water is to clean the dishes. Indeed, as listed in Table 2.2, dish–Stirling systems require just 20 gallons of water per MWh. According to a recent press release, SES claims that they would be able to produce electricity for 500,000 homes, for the amount of water consumed by just 33.

Linear Fresnel reflectors (LFRs)

Like parabolic troughs, these systems use a linear central receiver mounted above a mirror. Unlike parabolic troughs, however, the mirror and the receiver are separated. The receiver is fixed and several rows of flat or lightly curved mirrors are mounted on tracking stations to either side (Figure 2.7). In some cases there may also be a small parabolic mirror mounted above the receiver to further concentrate the solar energy reflected from beneath.

The potential advantage of the Fresnel system is in its simplicity of design. Since the receiver does not move, there is no need for complex fluid coupling units. Similarly, the mirrors can be arranged to focus on different receivers throughout the day, theoretically allowing them to be more closely packed together, saving space – an important environmental concern (see Chapter 7).

To date only small and experimental Fresnel systems have been developed, including a 1.5 MW system constructed by Ausra in Australia, a 5 MW

Figure 2.7 Linear Fresnel reflectors combine attributes of power towers and parabolic troughs (source: Areva Solar via NREL).

system in California and a 1.5 MW demonstration system by Solarmundo in Belgium. Further prototypes have been constructed at the German Aerospace Centre (DLR) and the first moves are under way to develop a 10 MW installation in central Spain.

The real breakthrough for LFR technology came in 2010 when the French engineering giant AREVA announced that it would be investing up to €500 million in the technology, beginning with the acquisition of Ausra. Plans are currently under way for a 500 MW facility in Australia. According to the man responsible for purchasing Ausra – Anil Srivastava – one of the main benefits of LFR technology is its relative simplicity, with its lack of moving parts. He also believes that there is real potential for cost reductions by localising manufacture in the countries where the systems will be used, particularly in emerging economies like India and North Africa.[28]

Solar chimneys or solar updraft towers

Solar chimneys (sometimes called updraft towers) are the odd one out of CSP technologies, and it is not obvious where to place them. Indeed, some commentators felt that they should not be included in this review at all. They do not use mirrors to focus the sun's energy, nor do they use a heat-driven turbine. Originally designed by the German engineering company Slaich Bergmann, solar chimneys resemble a large funnel placed upside down on the ground with the neck pointing up. The wide part of the funnel acts as a giant greenhouse. As the air in the greenhouse heats up it rushes up the central chimney, driving one or more turbines and generating electricity. In a sense, a solar chimney is more like a vertical wind farm, where the wind is created using the sun. The higher the solar chimney, the greater the pressure differential between the top and the bottom, increasing the power to the turbines.

Figure 2.8 A prototype solar chimney or updraft tower in Spain (source: Slaich Bergmann).

The first experimental solar chimney was developed in the early 1980s in Spain with funding from the German government (Figure 2.8). This was a 50 kW facility with a maximum height of 195 metres, while the collector covered around 46,000 m². Based on the experience of this tower it was estimated that a 100 MW facility would require a tower 1000 metres tall with a collector area of 20 km². This is obviously a massive area of land for a 100 MW power station and unless this can be dramatically reduced it seems unlikely that the solar chimney will find wide application. It does, however, have some possible advantages, including a relatively low tech and low cost design. Estimates for the cost of power supplied are also very low, with Slaich Bergmann estimating €0.07 / kWh.[29]

Over the last few years a number of plans have been put forward by a company called Enviromission for solar chimney projects, notably in Australia. To date though no specific progress appears to have been made. Other possible projects have been touted for Namibia, which in 2008 approved a 1.5 km high tower, with a base of 37 km² (3700 hectares). Called Greentower, this would not only have a capacity of 400 MW, but the base would also provide a controlled environment for growing cash crops. Should it be built, it would be the tallest man-made structure in the world.[30]

It should be pointed out that solar chimneys have attracted a fair amount of scorn from some quarters, with many seeing them as intrinsically unrealistic and ineffective. Nonetheless they still turn up in reports from time to time and it useful to know what they are.

Photovoltaic technology

Photovoltaic, or PV, technologies use sunlight to produce electricity using the photovoltaic effect. This is the term given to the ability of photons to 'push' electrons into a higher energy upon collision, knocking them out of their 'position' creating a separation of electrical charge and thus a difference in electrical potential. In practice this means exposing certain kinds of photo-sensitive material, such as silicon, to light to produce a direct current. The photovoltaic effect was first discovered in the early nineteenth century by the French scientist Becquerel, with the first functioning photovoltaic cell constructed by Charles Fritts in 1883 using selenium as the semiconductor. Just over 70 years later, the first practical photovoltaic cells were developed at California's famous Bell Laboratories in 1954. Because of their high price, PV was originally confined to providing electricity for satellites and probes, first powering the Vanguard I spacecraft. As a result, photovoltaics may be among the first human technologies that will be encountered by alien civilisations. PV continues to find use in space, but from the 1970s it began to be used widely in off-grid power applications. In the last 15 years the industry has undergone a revolution, with grid-connected installations growing at an average of nearly 25 per cent per year, spurred initially by support measures in Germany and Japan, and later the USA, Spain and now Italy and other Mediterranean countries. By the end of 2010, there were nearly 40 GW of PV installed worldwide, with around 16 GW installed in that year alone, up from 7.2 GW in 2009.[31]

There is a lot of terminology associated with solar PV, but the bare basics are fairly straightforward. The smallest unit of photovoltaic material is called a cell (solar cell). This contains the photovoltaic material along with positive and negative contacts or layers. Individual cells are then arranged into chains or series. In most cases PV cells are fragile and easily damaged (with the exception of some thin-film materials which can be deposited onto flexible plastic layers), so they are encapsulated into protective panels that can produce a usable output. For example fifty 3 W cells may be combined to produce a 150 W panel. A series of panels are then connected in an array. Ten 150 W panels would give a typical 1.5 kW domestic system. In domestic, grid-connected applications the current produced by the array is then passed through an inverter to convert the direct current (DC) into alternating current (AC) used by most appliances and grid networks.

As with CSP, there are a large number of different PV technologies, the principal differences being the various types of photo-sensitive material used, and the way the cells are interlinked. Different substances absorb different segments of the spectrum (bandwidth), and produce different voltages or have different conductive properties. With the constant pressure to reduce prices, there are also continual efforts to find the most cost-effective balance between cell and module efficiency and manufacturing and material costs.

Unlike CSP, photovoltaics do not require heat to generate steam and drive turbines. Because of this, PV systems need no moving parts (except in some cases for tracking systems which keep the panels pointing at the sun) or bulky generators. This makes them ideal for distributed and decentralised power generation, and indeed they have so far found great success in domestic and rooftop applications in the order of 1–5 kWp. Nonetheless a number of multi-MW installations have been constructed in the USA, Spain and Germany and there are plans for much larger installations in North Africa and the Middle East.

While PV and CSP are rivals in the current plans for utility driven GW-scale electricity generation, to date PV accounts for the vast majority of solar electricity installed, both centralised and decentralised. By the end of 2009, there were around 23 GW of solar PV installed worldwide, with the main markets in Germany, Spain, the USA and Japan, while by the end of 2010, this had reached 40 GW. In fact, in some countries PV now generates significant quantities of the electricity mix. In 2010, Spain produced nearly 3 per cent of its national electricity demand from solar, mostly PV, while a further 16 per cent was produced by wind.[32] Given that in 2010 Spain was Europe's fifth, and the world's tenth, largest economy, this is an impressive achievement, which despite the country's recent economic problems has been achieved in just a few years.

The two main 'families' of PV are crystalline silicon and thin-film, each of which has a number of subdivisions. Note that thin-film systems may use silicon as the semiconductor layer, but in this case they use a very thin layer, either crystalline or amorphous deposition.

Since the inception of the PV industry, crystalline silicon photovoltaics have remained dominant and account for the vast majority of installations to date. In the last few years, however, while the quantity of thin-film cells has expanded rapidly, its proportion of total installation has remained fairly static at about 22 per cent and may now be falling in the face of surging polysilicon production in the Far East. Nonetheless many continue to see thin-film systems (part of a suite of so-called second generation PV technologies) as having the greatest long-term potential for cost reductions in the solar market.

Crystalline silicon

Crystalline silicon modules consist of a number of thin, and extremely fragile, sheets of silicon (usually square or circular) arranged in grids. Each cell is connected to its neighbours to form a chain. A number of these cells are then encased in a module. This usually involves attaching the cells to a layer of glass and a polymer backing, and then encasing them in a metal frame, making the modules much more robust than the cells alone (Figure 2.9).

Figure 2.9 Crystalline PV in use in Villar de Cañas in Spain. In colour it would look blue (source: Yingli Green Energy).

Crystalline silicon photovoltaics can in turn be divided into two broad kinds – those made from monocrystalline silicon (i.e. a single large crystal sliced up) and those made from polycrystalline silicon (an ingot of silicon made of many crystals). This distinction comes from the way in which silicon ingots are grown and cooled and each has its own benefits. Monosilicon tends to have a more efficient rate of energy conversion, with modern cells achieving rates of around 25 per cent (up to 27 per cent under concentration)[33] and module efficiency typically around 16–20 per cent.[34] Polysilicon generally achieves lower conversion rates (module efficiency around 15 per cent)[35] but it is cheaper to manufacture and more plentiful, lowering the costs of producing the modules. Since silicon accounts for a large percentage of the cost of the cell, this can be a significant factor and recent drops in the price of polysilicon could revolutionise the sector.

When photovoltaics was still a small industry it was able to source its silicon from the 'off-cuts' of the semiconductor industry. Thanks to its rapid growth, however, by 2006 the industry was experiencing a major shortage of suitable silicon, prompting many manufacturers and researchers to look for alternatives and means of reducing costs. While the silicon shortage is now over, thanks to increased investment in crystalline silicon production, efforts to reduce costs and raw materials are still continuing. In fact at the time of writing the price of polysilicon has recently fallen dramatically, reportedly by up to 90 per cent since 2007.[36] It is unclear what the knock-on effect of this will be on the module price, and if it can be maintained as the global economy begins to recover.

Thin-film

Thin-film is the name given to those PV technologies in which the solar cell is composed of a thin layer (typically 2–300 μm) of photovoltaic material. Because of the thin layer of the deposition and the variety of materials which can be used as a substrate, thin-film technologies have tremendous potential for mass production and versatility, particularly since some can be backed onto flexible materials such as plastic sheeting. While the proportion of thin-film technology in the global market reached around 22 per cent in 2008, it had until recently played little role in the plans for large-scale energy generation in the desert, instead finding application in rooftop installations and portable chargers. This is all changing, however, as increasing conversion efficiencies and multiple band gap technologies mean that thin-film is becoming more commercially attractive. At the same time production of some has increased dramatically, with the manufacturing capacity of CIS cells expected to rise by 1.3 GW by 2012 in the USA alone, while thin-film manufacturer First Solar is one of the largest PV producers in the world.

In reality there is no one kind of thin-film solar cell. Various researchers and companies have used a wide range of materials deposited in a number of ways, often combining more than one layer of photovoltaic material. Although by no means exhaustive, the principal thin-film categories arranged by the type of photovoltaic are classed as:

- copper–indium–gallium–diselenide (CIS or CIGS)
- III–V cells such as gallium arsenide, named after the columns on the periodic table their constituent elements are found in
- cadmium telluride (CdTe)
- amorphous silicon (a-Si)
- dye-sensitised solar cells (DSC) and organic cells.

Amorphous silicon (a-Si) and other thin-film silicon cells

Thin-film silicon cells were among the first thin-film technologies to enter mass production and are produced by depositing a thin layer of silicon onto a substrate (glass, metal or plastic) that has been coated with a conductive layer. This coating is normally achieved using vapour deposition techniques, although other methods such as sputtering are also being explored.

For many years a-Si was considered rather inefficient and unstable, and was only able to achieve conversion efficiencies of around 4–7 per cent, although more recent systems have reportedly been able to achieve cell efficiencies as high as 10 per cent.[37]

Interestingly since different types of thin-film silicon have different band gaps (i.e. they absorb light more strongly at different wavelengths), new types of cells combining different silicon materials are being developed and are known as tandem cells. Amorphous silicon has a higher band-gap than crystalline silicon, meaning it converts visible sunlight into electricity more efficiently. By comparison crystalline silicon (c-Si) is at its most efficient at infrared wavelengths. By combining a layer of amorphous silicon with a very thin layer of polycrystalline silicon, absorption at both visible and infrared wavelengths can be optimised, producing cells with efficiencies of around 13 per cent.[38] Some systems take this further, combining multiple layers of photovoltaic material, each with a slightly different band-gap. These are known as multi-junction cells.

The principle manufacturers of amorphous silicon technology include United Solar Ovonics in the USA. In the coming years, many analysts expect production of a-Si panels to move to new facilities in the Far East, following the same path as crystalline silicon systems.

Cadmium telluride

In the last few years cadmium telluride or CdTe has enjoyed explosive growth as a form of thin-film technology. Indeed, one of the main proponents of this technology – First Solar – was for a while the largest single producer of photovoltaic cells in the world, producing around 1200 MW in 2009 at plants in the USA, Germany and Malaysia. Cadmium telluride is one of the few thin-film technologies that has so far been able to directly compete with crystalline PV and is of interest for this book, as it is currently being used in a number of large multi-MW installations, such as the 40 MW Waldpolenz facility in Germany, several 10 MW installations in China and a possible 2000 MW facility in the Gobi Desert (Figure 2.10).

The use of CdTe as a photovoltaic material goes back to the 1950s when it was realised that the compound has a band-gap which allowed efficient conversion of solar radiation. Early pioneers in the 1980s such as Kodak and Matsushita managed to produce cells with an efficiency of 10 per cent using a layer of CdTe backed onto another compound, cadmium sulphide (CdS). Further research showed that thinning the layers and adding conductive oxides could increase efficiency further, with a 15 per cent cell produced in the early 1990s, and a 17 per cent cell around 2010.[39] Nonetheless current commercial module efficiencies are normally much lower, with the best now at around 10–12 per cent.[40]

Current champions of CdTe technology include the previously mentioned First Solar, along with Abound Solar and Solexant. The Germany company Q-Cells also has a subsidiary called Calyxo that manufactures CdTe technology. As with some other forms of thin-film PV there is great potential for cost reductions in CdTe technology, both through improved industrial processes and also from the flexibility that could come with mounting it to flexible substrates, such as plastic sheets.

Figure 2.10 Thin-film CdTe photovoltaics in use in Canada (source: First Solar).

CdTe technology is not without its difficulties though. Aside from potential resource bottlenecks, which are mentioned in Chapter 7, cadmium metal is highly toxic, with potential health consequences for the manufacture and disposal of CdTe cells. While cadmium is stable when bound with tellurium, and has been shown to be stable in fire tests, it is clear that there will need to be a robust collection and recycling policy to ensure that end-of-life panels cannot cause an environmental hazard. The quantities of cadmium involved should not be overestimated though, and are still relatively small. Standard coal power stations have been shown to produce many more times the amount of cadmium per unit of energy compared with CdTe solar panels (see Chapter 7).[41]

CIS / CIGS cells

Copper–indium–gallium–diselenide cells are so called because they use a photovoltaic material composed of the elements copper, indium, gallium and selenium. While some thin-film cells in this category employ all four elements, others are restricted to just copper, indium and selenium. This can cause confusion over the different acronyms since the entire class of products is often simply termed CIS.

As with the other forms of PV, a great deal of research and development has gone in to creating the highest efficiency cell possible, with the current record for a CIGS cell standing at around 20 per cent set by a team at NREL in the USA.[42] This level of conversion efficiency compares favourably with most polycrystalline silicon cells (although it is still some way behind monocrystalline). More recent testing at the Fraunhofer ISE research institute in the German town of Freiburg has demonstrated complete module efficiencies of around 13 per cent.[43]

Principal manufacturers of CIGS cells include Solibro, a subsidiary of German company Q-Cells along with Solyndra (now bust), Nanosolar, MiaSole, SulfaCell and Frontier Solar. Although total global production in 2009 was estimated to be just 43 MW,[44] this is now changing rapidly with a host of new plant coming on line from 2010 onwards. By 2012, there may be 1.3 GW of production in the USA alone, with another 900 MW in Japan. In common with some of the other thin-film solar materials, there are questions over the economic availability of rare metals such as indium, and the possible effect this will have on the price of these technologies as they achieve greater scales (see Chapter 7).

Dye-sensitised and organic solar cells

As someone with a background in biology, I often used to wonder why those involved in solar energy were not making cells using processes similar to photosynthesis, a reaction pathway that has been tested and refined over hundreds of millions of years as plants have turned sunlight into fuel. Of course it turns out that I was wrong and that a great many scientists had indeed been exploring this avenue of thought, one manifestation of which is the dye-sensitised cell.

Most famously associated with Prof. Michael Grätzel and his colleagues at Switzerland's Ecole Polytechnique Federal de Lausanne, dye-sensitised cells use a photo-electric dye, a little bit like chlorophyll, attached to a layer of titanium oxide and immersed in an electrolyte solution in the presence of a platinum catalyst. This is then sandwiched between a positive and a negative electrode, the top one of which will be transparent to allow light to enter. Whereas in normal solar cells (of whatever thickness) electrons 'move' back and forward between a positive and a negative layer as light strikes them, in a dye-sensitised cell the sunlight striking the dye causes electrons to be excited and flow towards the transparent electrode where a charge is generated. In order to prevent the dye from degrading, an unexcited electron is reintroduced to the dye via the electrode on the back of the cell from the external circuit.

Currently, maximum efficiency for dye-sensitised solar cells is about 11 per cent,[45] although work is under way to test different dyes and combinations of dyes to achieve higher conversion rates or to expand the band-gap of the dye solution. Practical advantages which have been suggested for these types of cells include potentially low production costs, flexibility of application and the ability to operate under extremely low light conditions and higher temperatures. Whereas most PV cells have a minimum amount of light they will operate in, with dye-sensitised cells this is very low, and it has even been suggested that they could be used indoors, constantly generating low levels of electricity. As might be expected, however, these new materials are not without their drawbacks, in particular sensitivity to temperature and degradation. The liquid dyes can freeze and expand under cold and hot temperatures, potentially damaging the cell and stopping generation. Furthermore, at present the electrolytes that are used damage plastics, preventing application to flexible polymer membranes. Nonetheless this remains a promising area of research and in 2010 Prof. Grätzel was awarded Finland's Millennium Technology Prize in recognition of his work.[46]

Companies that currently supply dye-sensitised cells include G24i, Dyesol, Konarka, SolarPrint, Hydrogen Solar and Sony. Of these, G24i, which is based in California but with manufacturing facilities in the UK and Switzerland, has been among the first to offer a commercial product, and in 2009 provided a shipment of small panels to a Hong Kong based company for integration into bags.

GaAs cells

Made of the elements gallium and arsenic, GaAs cells were amongst the earliest PV cells to be manufactured, and are still used in applications where high efficiency is desirable, such as space or concentrator systems. GaAs can be monocrystalline, or thin-film crystal, with cell efficiencies recorded at 26 per cent and 28 per cent respectively.[47] Due to their high costs, GaAs cells are most favoured in concentrator applications (where the sun is focused onto a small piece of photovoltaic material), and efficiencies here have reached over 29 per cent.[48] Gallium and arsenic also find application in very high efficiency multi-junction cells, sometimes called III–V cells.[49] In these applications layers of GaAs or other materials (such as GaInAs) may be combined in layers with silicon. When used with concentrators, cells of this type have been able reach efficiencies of over 40 per cent.

Resource constraints

As thin-film technologies have made the leap from small scale and experimental to large industrial technologies, increasing numbers of questions have been asked about the availability of some of the rare materials they contain, such as indium or tellurium. Indeed, the availability of these materials has already begun to have an impact, with tellurium increasing rapidly in price as First Solar expanded its solar production line. In the last couple of years, questions have been raised about the availability of indium and gallium too as CIGS production has expanded, and their use in flat panel (not just LED panels) television screens has soared. While it is easy to overplay possible shortages, short-term bottlenecks in supply could have an impact on the economics of thin-film technology. This topic is discussed in more detail in Chapter 7.

Figure 2.11 Concentrating photovoltaic systems in Puertollano, Spain (source: Soitec).

Concentrating PV

Not to be confused with Concentrating Solar Power (CSP) or Concentrating Solar Thermal Power, Concentrating Photovoltaics (CPV) systems use mirrors or lenses to focus the sun's light onto a small area of photovoltaic material (Figure 2.11). By focusing the sun's light up to 1000 times (known as 1000 suns) the idea is to reduce the amount of expensive semiconductor material that is needed to produce a usable quantity of energy, and so reduce the costs of the system.

The principle of CPV is quite straightforward. In standard 'flat-plate' PV modules, a large area of photovoltaic material (crystalline silicon or thin-film) is exposed to the maximum naturally occurring sunlight. Normally, that maximum is achieved by installing the modules at an incline optimised for the latitude, but sometimes they are installed on trackers that can follow the sun as it passes across the sky. The amount of light that falls on a cloudless day (this varies according to location and season) is regarded as one 'sun', which is defined as 1000 W/m^2. In contrast, concentrating PV systems use lenses or mirrors to focus sunlight onto a small amount of photovoltaic material. This may be a series of reflective mirrors, a parabolic dish or trough or a lens. (Often a Fresnel lens is used, a flat lens that uses a miniature zigzag or sawtooth design to focus incoming light. When the teeth are arranged in concentric circles, light is focused at a central point. When the teeth run in straight rows, the lenses act as line-focusing concentrators.) The concentration ratio can vary: if the light that falls on 100 cm^2 is focused onto 1 cm^2 of PV material, the ratio is considered as 100 suns. If the light from 10 cm^2 is focused onto that 1 cm^2, the ratio is 10 suns, and so on. If the concentrated sunlight falls onto a well-designed CPV cell, the cell will produce at least 10 times, or 100 times, the electricity. In fact the conversion efficiency of solar cells increases under concentrated light, so the correlation can be greater than one-to-one, depending on the design of the solar cell and the material used to make it. At present CPV systems are divided into three broad categories: low (< 10 ×), medium (up to around 150 ×) and high (> 200 ×) concentration. Most of the utility-scale commercial systems discussed in this review operate at high concentrations. In the future, even higher concentrations are likely to become increasingly common. Importantly, since CPV can only operate efficiently in direct sunlight they require tracking systems to make sure they stay focused on the sun throughout the day.

At low concentrations, usually with single-axis trackers, CPV generally makes use of silicon cells. However, as mentioned above, concentrating PV also offers the option of shifting away from crystalline silicon to use the very high efficiency non-silicon cells. These multi-junction III–V cells (which use elements from columns III and V of the periodic table, typically gallium and arsenide) are prohibitively expensive for extensive use in large flat panel arrays. Concentrator systems, however, because they require far smaller and fewer cells, can afford the higher cost of multi-junction cells and yet still be manufactured at an acceptable dollar-per-watt cost. Indeed, combined with dual-axis tracking, the high efficiencies of these cells (typically 35–40 per cent) can enable the CPV units to operate at high module efficiencies, up to 27 per cent (compared with a module efficiency of 20 per cent for crystalline silicon or 12 per cent for cadmium telluride). This can greatly help reduce the cost per kWh produced.

At present there are several main manufacturers of III–V solar cells for use in CPV installations. Perhaps the most widely used is Spectrolab, a division of Boeing, and a company that regularly sets records for cell efficiency. Other companies include Emcore, Azur Space and JDSU. Concentrix also manufactures its own cells. Aside from efficiency, there are a number of other factors which module manufacturers might take into account when sourcing cells for their systems. CPV is still a relatively new technology, and as such there is an element of risk attached to deployments. Companies may be wary of using cells from smaller companies for fear that they might not be able to honour warranties in the event of a major programme. Bigger companies, with large backers, therefore have an advantage in terms of 'bankability' or risk reduction.

An important feature of CPV is its lack of water use. While it is important to cool CPV systems (the efficiency of a solar cell decreases with rising temperature) this is usually achieved by backing the cell onto a highly conductive metal, such as copper. Providing that this layer is sufficiently large, it is usually sufficient to keep the system cool. If additional cooling is needed, air-cooling systems can be employed. Given the pressure on water in many arid areas of the desert, and the cooling demands of parabolic trough solar thermal systems in particular, the ability to dry cool as standard could be an additional advantage in the future fortunes of CPV, as compared with concentrating solar thermal.

Land use is something that many CPV manufacturers are also keen to emphasise. Per GWh, high-concentrating CPV uses 2.2 acres (0.9 ha) of land compared with 4 acres (1.6 ha) for cadmium telluride thin-film and 4.2 acres (1.7 ha) for a solar thermal power tower. While this could provide an advantage in terms of permitting an environmental clearance, so far it appears to have little impact on the economics of these systems, since land costs typically represent such a small part of the overall system costs. While this is certainly true in the Western United States, it may not be quite so inconsequential in Southern Europe where land costs are higher and there is far less space of any kind available. In India too, with its massive overcrowding and demands on space of all kinds, one could imagine this being a positive advantage in developing large-scale solar systems.

So far CPV production has remained low, and installations number just a few tens of megawatts. Nonetheless it is in large-scale ground-mounted tracking systems that CPV could really be competitive and for this reason it is likely to be an important consideration in the plans for energy from the desert. At the time of writing there are 689 MW of concentrating photovoltaics under construction and in various stages of development around the world – overwhelmingly in the USA, but also in Spain, Portugal, Mexico and Australia (of this 689 MW, some 414 MW have signed a power purchase agreement or been signed up to a feed-in tariff, so these are perhaps more definite). There are also a number of potential large-scale projects in India and South Africa, although these are still uncertain.[50]

PV tracking systems

Tracking systems are widely used for PV applications in areas with good solar resource, and allow the panels to be continuously moved throughout the day to face the sun. In areas with less sunshine the increased efficiency may not be enough to offset the increased costs of the installation. In large desert-based applications, tracking systems will almost certainly be widely used. There are some notable drawbacks associated with tracking systems, however. To begin with they introduce moving parts to an otherwise rigid system. This increases the risk of technical failure and subsequent downtime. Secondly, panels must be spaced further apart to ensure that they do not shade each other, increasing the necessary land area of the solar installation.

For concentrating PV (CPV) applications, tracking systems are essential.

Solar energy for desalination

While we have so far looked primarily at the idea of electricity from the desert, it is important to realise that there are many uses for both PV and CSP that would not necessarily involve the power they produce being fed directly into the grid. In some instances the electricity could be used on-site for industrial applications, and one of the most likely and important uses could be in desalination.

Globally, demand for clean water is growing, and many countries in the world suffer acute water stress (broadly defined as less than 1000 m³ per capita, per annum). While conservation and rational use of water should be the first step towards tackling this, desalination is growing in popularity, with many coastal cities now considering it a cheaper alternative to transporting clean water from inland.

Desalination plants work by taking in sea water and removing the salt. This process is very energy intensive and not at all environmentally friendly. As a by-product of the desalination process they produce large quantities of warm brine that must be disposed of. In addition they often use harmful chemicals, such as anti-foaming or anti-scaling agents, or biocides to kill marine organisms. These are unpleasant chemicals that can have serious impacts on our already stressed marine environment.

It is obvious though that many of the countries that suffer the most severe water shortages also have abundant sunshine. The Middle East for example accounts for around 75 per cent of all the desalination worldwide and has a growing need for fresh water. By using renewable solar electricity to power desalination plants there is potential for some of the negative impacts of desalination to be mitigated. Currently there are three main methods of desalination: reverse osmosis (RO), multi-effect desalination (MED) and multi-stage flash desalination (MSF). In theory, solar energy can be used to power any of these three systems by providing the electricity; however, in practice some systems are likely to be more suitable than others. Multi-stage flash desalination has been left out of several of the long-term studies due to its high energy consumption, with the general assumption being that it is not a sustainable system. MED systems seem particularly suitable for CSP installations as they rely on both heat and electricity to produce clean water, which CSP stations produce in abundance. They also have a higher thermal efficiency than flash desalination. Reverse osmosis systems, which use membranes to filter out salts from the water, only require electricity to pressurise the water and so can be fed by any source of energy, renewable or otherwise.[51]

Using CSP for desalination is a major component of the DESERTEC Initiative and more information can be found in Chapter 5.

Solar energy for hydrogen

Providing transport fuel represents a key challenge for the future. One of the technologies being considered for greener transport is the hydrogen powered fuel cell. These use hydrogen to produce electricity in the vehicle itself, by reacting it with oxygen in the presence of an electrode. In theory the only by-product would be pure water. Of course the hydrogen must be produced and transported to the car, and the means by which this is done has a major bearing on the 'green-ness' or otherwise or the hydrogen fuel cell. Currently hydrogen is usually produced through reforming methane, in which the gas is split into hydrogen and carbon dioxide. While this is an efficient process (around 80 per cent), it is not clear that this is an appropriate and sensible use of natural gas, particularly given that natural gas can be used to power vehicles directly in a compressed or liquified natural gas engine.

In order to make the hydrogen fuel cell truly environmentally friendly, new low-carbon methods of hydrogen production are required. Fortunately it is possible to produce hydrogen by using an electrical current to split water in the presence of a catalyst. Assuming that this electricity is produced by renewable means it offers the potential for a really low-carbon transport fuel, and one that would be theoretically limitless. Both large-scale PV and CSP are being actively considered for providing on-site electricity for hydrogen production.

Efficiency losses in hydrogen manufacture can be significant, however, and when combined with the losses of efficiency in a fuel cell, and the energy required to transport the hydrogen, can mount up. In its report on energy transmission in the Mediterranean region, the German Aerospace Centre (DLR) estimated that these losses could amount to up to 75 per cent over a distance of 3000 km, and so discounted hydrogen as a means of transportation.[52]

In the long run it remains to be seen whether hydrogen, natural gas, biogas, electricity or some other fuel will be powering our transport in the future, but cheap and clean electricity from a future solar power source would be a strong contender – whether it is used for hydrogen production or directly for plug-in vehicles.

Energy transmission and distribution

One of the crucial technological hurdles which must be overcome to truly utilise large-scale energy from the desert is the availability of grid connection. As has been mentioned desert areas are often sparsely populated and far from the places where electricity is most needed. To transport the electricity to the load points will require the expansion of the grid system. Put simply there are two main kinds of grid electrical transmission: alternating current (AC) and direct current (DC). While AC is the standard form used in the domestic grids of Europe, North America and Asia, DC is the one that is of most interest for power transportation. There are a number of major reasons for this. The first is that in some regions, such as Europe, the continental grid is divided into many different zones. While each of these uses AC, the grids are not synchronised with each other and so cannot easily exchange electricity. This would be equally true of energy coming from the North Africa and Middle East region, as envisaged in the DESERTEC project.

Perhaps more important though is that over long distances, DC transmission at high voltage suffers much lower energy losses than AC current at high

Table 2.4 Costs of HVAC and HVDC transmission

	HVAC (750 kV)	HVAC (1150 kV)	HVDC (600 kV)	HVDC (800 kV)
Overhead line cost (million €/1000 km)	400–750	1000	400–450	250–300
Sea cable cost (million €/1000 km)	3200	5900	2500	1800
Terminal cost (million €/1000 km)	80	80	250–350	250–350

Source: Robert Pitz-Paal, ECOSTAR – European Concentrated Solar Thermal Road Mapping; http://www.vgb.org/data/vgborb_/Forschung/roadmap252.pdf

voltages. For overhead lines, HVAC suffers losses of 8 per cent per 1000 km, compared with just 2.5–3 per cent per 1000 km for HVDC. For underwater cables the difference is far more dramatic; HVAC losses from sea cables can be as high as 60 per cent per 100 km, compared with 0.25 per cent per 100 km for HVDC! Clearly there would be no point in generating electricity from the desert or offshore, only to see it lost in subsea transmission.[53] HVAC does have advantages over short distances, with power losses from terminals significantly less than HVDC stations (0.2 per cent per station compared with 0.6 per cent per station), but at distances of over 30 km or underwater, HVDC is the technology of choice.[54] There are other advantages to direct current too, as more energy can be transported for a given number of pylons. One study shows that to transport 10 GW of electricity using HVAC lines would require five rows of pylons, with a combined width of 425 metres. With HVDC only two would be required, with a width of 100 metres.[55] This is important, because transmission lines are costly and environmentally damaging – they are also, quite rightly, difficult to get permission for.

Looking to the future, other transmission technologies are being developed, for example super-conducting DC. In these technologies liquid nitrogen is used to cool the transmission lines,[56] reducing resistance and leading to very low transmission losses (reported to be up to 150 per cent lower than standard AC or HVDC).[57] While energy is used in the refrigeration of the lines, this is less than the energy saved from the transmission losses, and the cables and refrigeration can be integrated into a pipeline type structure. Aside from the potential to reduce transmission losses, super-conducting DC can transport large amounts of power with a small thickness, or cross-section, of cable, which again has potential advantages. At present it is not clear over what distance super-conducting DC will be cost effective, and with the technology still in the early commercial phase it seems likely that HVDC will remain the technology of choice for the next few years at least.

Despite its advantages, developing HVDC requires significant capital expenditure and could represent a major upfront cost in large-scale renewable energy development (Table 2.4). Part of the reason for its extra cost is the need to have substations at the beginning and end of the line to convert the power back into usable AC current, but it is also the case that building any new grid infrastructure is costly, and in many cases prohibitive for individual developers and power producers. Because of this, systems will need to be found which allow the costs of grid connection to be borne by many parties. One obvious measure is to cluster new projects to allow them to share common infrastructure. Another will be for governments and grid operators to take a strategic approach and develop the grid in certain key areas, encouraging the solar development to come to them. Already in Europe plans are under way for an offshore grid, which will allow speedier and cheaper connection for offshore wind and wave energy sources, and similar plans have been discussed for the deserts of North Africa. The technology itself is certainly nothing new, having been around at least since the 1950s, and there are existing HVDC lines all over the world. One cable linking Norway to the Netherlands is over 600 km long, of a similar scale to those that will be needed to connect Europe and North Africa, and reportedly paid for itself in less than two years! In any case, these kinds of links are certainly realistic and dozens will be necessary to realise the full potential of energy in the desert (see Chapter 6 for more information on DESERTEC).

Conclusion

Whenever I look into these things, I am always impressed by the number of areas and directions in which technology advances, and how seemingly unrelated discoveries can suddenly play a crucial role in the development of another field. In every direction – from mirrors and photovoltaic materials to new polymer substrates and software control mechanisms – the race is on to reduce the costs and improve the reliability of solar technologies. Already these have had impressive results and the price of solar electricity is set to fall rapidly in the coming years. I hope that this chapter will have given the reader some idea of the breadth of the sector and some knowledge of the different systems that are being employed, if only so they can spot glaring errors in press reports or are driven to go and look at these systems in more detail.

Chapter 3

Energy from the desert

Several years before beginning this book, I found myself standing on a rooftop looking out over the mud-brick oasis of Siwa, in the Western Desert of Egypt (Figure 3.1). Despite being only October, the sun beat down with an unending ferocity that had caused the town's population to take to their homes for shelter. The Egyptian desert is hot and the traditional houses in this part of the world are built accordingly, with thick mud-brick walls and small deep-set windows – cool in summer, warm in the chilly desert nights. The roof of the particular building I was on had a more modern addition, in the form of a small parabolic mirror, which I believe was designed to heat water. The combination of traditional architecture with solar water heating technology made the building – a hotel – one of the most comfortable houses in the oasis, far more appealing than the handful of modern apartment blocks which had sprung up around the edge of the town, complete with noisy air-conditioners. Beyond the houses were the date palm plantations, and beyond that, a vast sea of yellow and white that made up the desert. To the west were the endless dunes that stretched to Morocco and vast plains of gravel and fossilised seashells that spoke of a time when the whole area was submerged beneath the sea. I was struck by the thought that the union of the Berbers' age-old customs with new technology offered a solution to the development of the oasis, which is just beginning to show the scars of traffic, tourism and pollution. In the future, roof-top installations could provide hot water and electricity for the houses, while free-standing PV or wind turbines could pump water for the date palms or provide fuel for electric vehicles. Covering just a tiny portion of the desert outside the oasis with PV or concentrating solar would power the town and the nearby military bases and bring jobs to the community. It might also return to the people of Siwa something of the independence that living in the desert must once have offered. At the time of my visit I was not yet thinking of the potential for the desert to provide electricity for the big cities of Alexandria and Marsa Matruh on the Mediterranean coast, and perhaps Siwa is too far away, but the possibilities are there, a small example of what could be achieved under the right circumstances.

Figure 3.1 Satellite image of Siwa oasis in Egypt, 2009 (source: SSTL 2009, all rights reserved, supplied by DMCii).

Deserts of the world

The deserts of the world are vast. From the high Atacama in South America to the Gobi desert of Mongolia, they occupy nearly a third of the earth's land surface – and they are growing. Erosion, over-grazing, deforestation, salination and climate change are all causing a slow but steady expansion of the world's deserts. Already in Africa huge areas of agricultural land in the sub-Saharan zone, known as the Sahel, has been lost to the advance of the desert. By the late 1990s the UN was reporting that 73 per cent of drylands used for agriculture in Africa had been degraded by desertification.[1] In 1994, Niger alone was reported to be losing 2500 km².[2] In India and China deforestation, coupled with over-exploitation of land and water for agriculture and pastoralism, has spurred the advance of the Thar and Gobi deserts (although China's massive afforestation programmes may well reverse this). In Australia droughts and salination (caused by excess irrigation of dry agricultural lands, and the subsequent evaporation of the water) are causing the country's already vast deserts to grow even further.

As large as the desert areas are, so too is their potential for energy production. Every day the earth receives more energy in the form of solar radiation than human society currently uses in a decade. Harnessing the energy from just a fraction of the land available would provide abundant electricity, without the worry of carbon emissions, mining, air pollution or acid rain. The question is how to harvest this energy, how to get it to where it is needed, and how to pay for it.

As a student in school I vaguely remember being shown maps of the world with black squares dotted around them (some of them hovering ominously over Las Vegas). The black squares represented the theoretical amount of land area it would take to power the world, assuming we could harness all of the energy that fell on it. Fifteen years later and these black squares or circles can still be found, and a quick search of Google or Wikipedia still brings up examples. Now in 2010 these circles must have become a little larger, as the demand for energy has soared, but they have also become a little less hypothetical, and the race is on to realise the practical deployment of solar on these scales.

Deserts are not only large, they are sparsely populated and receive a tremendous amount of sunshine. Arid and hyper-arid areas encompass around 17,650,000 km^2, one-third of the land on the planet. Covering the whole of this area with currently available PV technology (10 per cent system efficiency and 50 per cent space factor) would generate 1.67×10^5 Mtoe of energy (6990 EJ) – more than 13 times current global energy consumption. By the same token, some calculations suggest that covering just 4 per cent of the available desert area (roughly the same size as the Gobi desert) in photovoltaics (15 per cent capacity factor this time) would be enough to generate around 412 EJ/yr, roughly equal to global primary energy demand from all sources for 2002 (or more than 5 times global electricity demand).[3] Since that calculation was conducted, global energy demand has increased to nearly 500 EJ/year (~12,000 Mtoe).[4] Fortunately, technology has also advanced, with large PV in the deserts achieving capacity factors of nearer 20 per cent and CSP installations of 25 per cent and above.[5]

Of course 4 per cent of the global desert area is still a truly enormous amount of land, and it is not clear whether solar energy on such a vast scale is currently feasible or even perhaps desirable. Indeed, this is where these sorts of examples fall down. To anyone who has spent any time in their midst, deserts are not just empty spaces on maps. They contain people, wildlife and areas of breath-taking natural beauty. The thought of obliterating the elegant dunes of the Sahara, the Mojave wilderness or the delicate oases of Egypt under an endless sea of solar installations is not appealing. At the same time there are a great many practical limitations to using the deserts as a primary energy source. In most cases they are located far from where the electricity will be needed, necessitating expensive energy transportation. There may be problems with providing sufficient process water for the systems, and for the communities that will be required to operate and maintain them. There are also concerns over security and ownership. Many desert parts of the world are politically unstable or brush against international borders, while some of the grandest plans envisage energy produced in one continent being 'piped' to another. How will these systems be regulated and protected, and will there be sufficient international good-will to maximise the benefits they could bring?

In reality it unlikely that large-scale solar will be called upon to meet total global demand any time soon, nor will it be concentrated in a single desert. Rather, large-scale solar power systems will be developed in clusters all around the world, from Mexico to China, Namibia to Morocco. They will be complementary to the other large-scale energy sources of the twenty-first century – offshore wind, domestic renewable and heating sources, marine and tidal, and perhaps nuclear and gas. Energy efficiency too must play a role. All energy has environmental consequences, so the less of it we need the better. What the following examples do is give some idea of the potential that is out there in the world's deserts, they are not in themselves representative of current plans or realities – although as we shall see, the situation is changing fast.

Availability of desert land

So what is the actual availability of desert land in the various parts of the world? Well, as mentioned, desert covers around a third of the land on earth, some 17 million square kilometres. Table 3.1 shows the area of the world's major deserts and the solar radiation they receive. As can be seen, all of the continents, with the exception of Europe, have significant desert areas. While the Mediterranean region receives ample sunshine, it cannot be classed as true desert, and is technically classed as a sub-humid zone. Nonetheless, the fact that Spain has emerged as a world leader in the deployment of both CSP and PV technologies shows that the possibilities for solar power are far wider than just pure desert regions, and depend on a host of other factors including energy prices and infrastructure and the political environment. So while this book takes as its focus the potential for energy production in the true deserts, the range of application for these technologies is far, far wider.

As important as the availability of land and the level of solar irradiation available, is the location of the desert area. Developing a grid to transport electricity is expensive and transmission can result in significant energy losses, so a piece of desert thousands of miles from the nearest major source of demand is less likely to be a prime candidate for large-scale solar energy, at least in the first incidence. Similarly, those areas of desert that are already near existing grid infrastructures may be easier and cheaper to develop, and because of this there are some desert areas that are of more immediate interest for solar energy than others.

We will now take a quick look at the major desert areas across the continents and give a brief description of their size and their suitability for large-scale solar energy.

North America

In North America deserts cover approximately 1.32 million km^2, largely in Northern Mexico and the South West of the United States. Of particular

Table 3.1 World deserts by size and insolation

Desert	Continent	Area in km²	Average insolation (MJ/m²/d)
Arabian	Asia / Middle East	2,330,000	22.24
Gobi	Asia	1,300,000	16.53
Kara Kum	Asia	350,000	16.34
Kyzyl Kum	Asia	300,000	16.34
Taklamakan	Asia	270,000	16.19
Kavir	Asia	260,000	18.33
Syrian	Asia / Middle East	260,000	18.10
Thar	Asia	200,000	21.44
Lut	Asia	50,000	21.09
Sahara	Africa	8,600,000	23.52
Kalahari	Africa	260,000	26.46
Namib	Africa	140,000	22.54
Great Basin	North America	490,000	20.32
Chihuahuan	North America	450,000	19.68

Source: K. Komoto, M. Ito, P. van der Vleuten, D. Faiman and K. Kurokawa (2007) *Energy from the Desert: Practical Proposals for Very Large Scale Photovoltaics*, Earthscan.

interest are the Sonoran and Mojave deserts, which combined cover approximately 380,000 km² and receive an annual solar irradiation of 17–21 MJ/m²/d. Large parts of these regions are relatively near to major load centres in Mexico, Nevada and Southern California and their proximity to well-developed grids, high-tech industry and agriculture makes them well placed to be among the first deserts to see large-scale deployment. Indeed, this is already beginning to happen, with nearly 10 GW of concentrating solar thermal energy under development in South Western USA, and roughly 2 GW of large-scale PV in the pipeline (the exact mix of technologies may fluctuate as developers opt for the most cost-effective technology at the time). The combined factors of high energy prices and growing demand in California, an outstanding research and business environment and an increasingly supportive political atmosphere, mean that this region of the United States is likely to continue to be a world leader in the development and deployment of this technology.

It is interesting to note how quickly the situation is progressing. A 2004 book, entitled *Energy from the Desert*,[6] envisaged Very Large Scale Solar as dealing with PV installations of 10 MW to 1 GW. At the time, all but the lowest end of this range seemed purely hypothetical, whereas just a few years later developments of such a scale have become the standard for large solar in many regions. While most of these involve CSP, there are more than 100 PV installations currently operational in the 10+ MW category, and in the USA there are PV plants currently under construction with capacities of over 600 MW (see Chapter 5).

Australia

In Australia there is plenty of desert, with the Great Sandy, Simpson and Great Victoria deserts alone making up 15 per cent of the country's land mass. Obviously this provides more than enough solar resource to supply the nation's domestic energy demands of 237 TWh.[7] Using the land-use figures supplied by DLR (Chapter 2), this implies that the country's total electricity needs could be met by 1700 km^2 covered in parabolic trough technology (at 8 m^2/Mwh/yr). The question therefore is whether this technology will be harnessed and what shape it might take. For example, the large distances between Australia's cities will likely mitigate against a single hub for solar energy, and will call for a number of separate solar hotspots. One can imagine individual populations like Perth and Sydney, as well as smaller towns, with their own dedicated solar power sources. Decentralised energy too could play a crucial role. In the future, Australia's potential for truly large-scale solar deployment – potentially moving beyond domestic demand – will depend on its ability to trade energy with its neighbours – something hampered by its geographic isolation and the scale of the country. On the face of it, Indonesia would appear a good target for electricity exports. It has a huge population and a rapidly growing economy, but the distances involved are vast, with the main population centres of Java and Sumatra more than 2000 km from the nearest part of Australia and even further from the desert regions. New Zealand too is over 3000 km away, across the rough southern ocean. Because of this, it seems likely that Australia's potential as a renewable energy exporter will be limited, or perhaps shifted to exporting energy in another form, such as hydrogen or other products produced from solar power.

Interestingly, Australia has the potential to use its solar energy resource to help green another environmentally destructive industry – mining. Worldwide, mining uses a huge amount of energy, and brings with it numerous environmental and social consequences, everything from water pollution and habitat destruction to forced migration and social conflict. At the same time it remains an essential enterprise to human society, providing critical raw materials. By developing large-scale solar to power aspects of its mining industry, Australia, and other countries, could help to mitigate at least one of the negative impacts of one of its most important industries, and in doing so reduce the carbon footprint of commodities such as gold, copper and zinc (solar would not of course, absolve the mining industry of all its other environmental and social responsibilities!). In fact Australia is in an excellent position to power almost its entire energy needs (fuel, transport, electricity) using solar power, should it choose to do so.

Asia

Asia is undoubtedly one of the most interesting locations for large-scale solar energy. Many Asian countries have growing populations, and the region as a whole is undergoing extraordinary economic development. This has led to a surge in energy consumption, the electricity part of which comes largely from coal, and a rapid rise in pollution. According to one survey by the World Bank, of the 20 most polluted cities in the world, 16 are in China[8] while several major cities in India regularly make similar lists. In some parts of China's coal belt, the sky and streets are permanently blackened with ash and soot. Cancer rates are higher than the national average, respiratory diseases are rife and the rivers and forests are heavily polluted. In 2007, China's State Environmental Protection Agency pronounced large sections of the Yangtze River to be 'biologically dead',[9] and a recent expedition to find the Yangtze River dolphin has drawn a blank. The species is presumably extinct in the wild.[10]

Emissions of greenhouse gases have also risen rapidly. In 2008, China overtook the United States as the largest single emitter of greenhouse gases, and its per capita emissions are now on a par, or even higher, than those of some European countries (such as Sweden).[11] In addition, acid rain and respiratory problems are serious issues in China and its neighbours. Some estimates suggest that as many as 750,000 people may die as a result of air pollution in China during a normal year.[12] The problem is also far from local. Satellite images show how the smog and particulates generated by China now spreads across the country and around the globe, reaching the west coast of the United States.[13] But China is also investing in renewable energy in a big way, and it does have important desert areas. The Gobi desert covers around 1.3 million km² split between China and Mongolia, while in the west the Takla Makan, the great barrier on the Eastern silk routes, spans over 270,000 km². Although these deserts are relatively far north, and receive only around 70 per cent of the irradiation of their counterparts in Arabia or North Africa, they still have vast power generation potential. Indeed because of the nature of the Chinese economy and political situation, they have some of the highest potentials for large-scale solar in the world. Covering the whole Gobi desert in PV would power the world (although it would make a lot of people and antelope unhappy!). Harnessing just 1 per cent (~13,000 km²) with PV could provide at least 1000 TWh, roughly one-third of China's 2008 electricity consumption of 3.48×10^3 TWh. If CSP technology were used, 1000 TWh could probably be produced with around 6000–7000 km² (at 8 m²/Mwh/yr) or less.

India too is growing rapidly, with its economy expanding by 7–9 per cent a year. Like China, much of its power comes from coal. Unlike China, however, India remains a largely rural country, with nearly 75 per cent of the population outside the urban areas (although perhaps not for long). Poor villagers and farmers continue to rely on timber and buffalo dung for heating and cooking fuel. This, coupled with a rising population and growing pressure on land, has led to increasing desertification and deforestation, bringing with it a devastating loss of biodiversity, health problems and falling standards of living for many of the rural poor. As these people shift to more efficient and available forms of heating it seems likely that fossil fuel consumption will soar in India, exacerbating climate change and the demand for oil, gas and coal. The deployment of solar power, at both a large scale, and at the village level, could be an important tool in overcoming this challenge. (This has actually been happening in the Indian subcontinent through the solar home system, which has brought light and power to hundreds of thousands in the region through small solar systems.)

Fortunately for advocates of big solar, India has large desert areas. Again, utilising just a small part of these areas for CSP or photovoltaic power stations could produce a significant proportion of the nation's needs. The Thar Desert covers around 210,000 km^2, split between India and Pakistan, and covering portions of the states of Gujarat, Rajasthan, Haryana and, in Pakistan, Sindh. PV or CSP deployed on roughly 5 per cent of the Thar Desert could technically generate around 1000 TWh of electricity, nearly twice as much as India's 2007 electricity demand. Deploying solar on 0.5 per cent (roughly 1000 km^2) of the Thar Desert area would therefore be capable of producing 20 per cent of India's electricity at today's levels. While this is still a large amount of land with consequences for humans and biodiversity, it must be put in context. Currently, coal mining in India covers thousands of square kilometres, with much of this land found in highly populated and wildlife rich areas of Central India, such as Bihar and Chattisgarh. In the Damodar Valley alone, coal mining affects around 4000 km^2, and possibly more.[14] Indeed many of its scheduled tribal peoples and some of India's most endangered wildlife, such as the tiger, live in areas threatened by coal mining and hydropower, something which the judicious deployment of solar power might be able to relieve. There is also surely a difference in the level of damage caused by a CSP or PV installation compared with coal mining. A few thousand metal and glass panels and their accompanying substations and pylons seems far more benign than endless tonnes of ash clogging up water courses, heavy metals in the ground water and hundreds of square kilometres of underground fires – the problems faced by many unfortunate enough to live in India's coal belt.

Middle East and North Africa (MENA)

Home to the great deserts of Arabia and the largest desert of all – the Sahara – the North Africa and Middle Eastern regions have the greatest solar energy potential of any area on earth. From Baghdad to the Atlantic coast of Morocco, deserts span nearly 5000 km from east to west, making the energy generating potential of the region almost endless.

To date this potential remains untapped, yet the energy demands of the Middle East and North African countries are set to nearly triple by the middle of the century, due to continued population growth and economic development. In addition, aside from the domestic demand there is tremendous potential for Middle Eastern and North Africa countries to export huge amounts of solar generated electricity to their neighbours, both in Europe and, potentially, in Sub-Saharan Africa.

In fact, the MENA countries are the location of one of the most ambitious and advanced proposals for solar power currently under development. The DESERTEC scheme, which is an initiative of European and North African industrialists, banks and political groups, would see European and MENA consortia build large-scale CSP plants across North Africa. Initially these would provide for the domestic MENA market, with the excess being exported to Europe (for more information on DESERTEC see Chapter 6). In the longer term it is estimated that this programme could provide up to 15 per cent of Europe's electricity.

There is potential too for these deserts to generate electricity for the power hungry nations of the Persian Gulf. The United Arab Emirates, along with the Sultanates of Qatar, Bahrain and Kuwait have the highest per capita energy requirements and carbon emissions of any nations on the planet. They also have a constant demand for fresh water, most of which must be supplied by desalination. The potential for solar power to play an important role in the Gulf region has not been ignored, and a number of encouraging developments are under way, most notably the MASDAR project in Abu Dhabi – an audacious attempt to develop a showcase 'eco-city' in the desert (see Chapter 6).

Perhaps the greatest challenge for the MENA region will be to overcome the issues of ownership and political insecurity. Who will own and benefit from the expansion of solar power in the region, and will it be able to overcome its historic (and on-going) instability long enough to allow it to develop?

Sub-Saharan Africa

Sub-Saharan Africa too will have a need for new energy sources as its economies and populations expand. While central Africa is not ideal for concentrating solar, due to its humid environment (less of a problem for PV), southern and eastern Africa do have massive potential. South Africa and Namibia both have desert areas of exceptionally high insolation in the Kalahari and Namib, which span 290,000 km^2 and 150,000 km^2 respectively. Aside from these true desert areas, much of the rest of the region is semi-arid, and, like similar regions in Spain and the USA, could be suitable for solar. Due to the low population density of much of Southern Africa, it is unlikely to be a major hub for large-scale centralised solar power deployment, although there is tremendous opportunity for countries to satisfy domestic demands with regional and distributed generation. The exception to this is likely to be South Africa. Indeed, recently South Africa has been increasing its energy deployment to meet the needs of its industrial and mining sectors, and to export to its neighbours. Unfortunately much of this new capacity has come from coal (including a highly controversial proposal from the World Bank to provide funding for a 3.3 GW coal power station, in the face of opposition from the USA and EU, the World Bank's main funders). Large-scale solar, either in South Africa or its neighbours, could provide an alternative in the future, and tentative plans are under way for the first pilot projects.

East Africa, while containing few areas of true desert, nonetheless has an enormous solar resource, and would be well placed to develop solar technologies. To date most of the talk about renewables in the region has focused on wind, and the current high costs of solar technology are likely to prove a barrier for several more years at least.

South America

In South America the true desert areas are confined to the Atacama in Chile (with a bit in Peru) and the Patagonian desert in Argentina. Despite its size (140,000 km^2) and high levels of insolation, the Atacama Desert seems an unlikely candidate to be an early mover in larger scale solar, thanks largely to its altitude and relative isolation. There are few large energy nodes nearby and taking energy to the more densely populated regions of South America would necessitate major transmission works. Nonetheless is seems likely that solar could still have a role to play in Chile, which imports more than 80 per cent of its energy needs. Whether this solar will be located in the desert regions, however, seems more uncertain. It is possible that solar could begin to play a role in providing energy for copper mining – the main industry in the Atacama – and there are some initial signs of this. In 2010, Codelco, the state-owned mining company, announced that it was planning to invest $50 million in solar power. If completed, this will provide energy for one of the company's principal sites, potentially cutting its oil use by 30 per cent.[15] This is just one of a number of solar plants being planned for the region.[16]

Far to the South of the Atacama, the Patagonian desert covers around 420,000 km^2, largely in Argentina. As with the Atacama, there are relatively few large electricity load points nearby. Export to Buenos Aires would be possible, but the distances involved are significant. Insolation in the Patagonian desert is also relatively low, around half of that found in the Sahara. It seems more likely that the needs of such a dispersed community as that found at the southern tip of South America could be better met through decentralised wind and solar installations, or through large-scale wind.

Chapter 4

Economic and policy aspects of solar power, and the status of regional markets

If the first question anyone asks about solar power is, 'Does it work?', the second question is usually, 'How much does it cost?' Currently, there is no simple answer to this. Each technology has a different cost, and in many cases each project will be different, depending on the local conditions. The level of solar resource in a region will have a heavy influence, particularly on PV, which is very sensitive to fluctuations in the insolation, and which can function in lower light conditions (as opposed to concentrating solar thermal technologies which generally require direct sunlight to generate electricity). A 1 kWp photovoltaic system in Northern Germany can be expected to produce 850 kWh per year,[1] while one in Southern California or India would generate around 1600–1800 kWh.[2, 3] For CSP this is a bit less of an issue, as these sorts of facilities can only be constructed in areas with a high level of direct sunshine, implying a lower range of solar conditions. Output can also be moderated with storage technologies and control systems.

Aside from solar resources, the availability of grid connection and the local policy environment too will have a major influence on the price of solar generation. The cost of finance is important, since the majority of expense in a solar system is in capital investment, as the fuel is free. All of these variables mean there is a wide range of cost estimates available. Anyone searching for the price of solar energy will discover no simple figures, but ranges of prices, some of them quite large.

Having said all that, the fact is that solar energy, particularly PV, is still quite expensive, although costs are dropping fast. Arguments can be made that if the true costs of fossil fuels were incorporated into their price then they would be prohibitive, and there is more on this below, but for the economic system that we currently have, solar technologies are still relatively pricey.

In 2008 the consultancy firm McKinsey estimated price ranges for PV power at $0.09–0.26 / kWh depending on technology, scale and location.[4] Solarbuzz for its part estimates the 2010 cost of solar electricity produced in the sunny part of the USA at $0.19–0.32 / kWh depending on the size of the system (the lower end is for large centralised systems > 500 kW, the high end for residential systems of 2 kW).[5]

In 2004, a study into the feasibility of very large scale PV came up with a range of cost estimates for 100 MW installations, based on varying insolation, module price, tracking technology and local wage costs. These findings are summarised in Table 4.1 and give a useful indication of the cost range for large-scale photovoltaics.

One thing that all the various groups are united on though, is that the costs of PV has fallen, and will continue to fall in the future. Since the 1980s, prices have fallen dramatically, dropping by around 20 per cent every time the installed capacity doubles (despite a short-term blip in 2003–2006 when prices actually rose and stagnated due to a bottleneck in the silicon supply). Nonetheless the overall trend is clear. In 1982, crystalline silicon PV modules cost $27/Wp,[6] falling to an average of around $3.29/Wp in 2010 on the US and European markets (modules tend to make up around half the cost of a PV system).[7] Looking at Table 4.1, this predicts a possible price range for large-scale solar in the desert of roughly $0.11–0.16/kWh, at $3/Wp, without government subsidy. Just before going to print at the end of 2011, the average module cost had fallen to $2.49/Wp in the USA, and was still dropping.[8]

These prices are expected to continue declining. McKinsey has estimated that the cost of crystalline silicon PV will fall by 5 per cent per year, and of thin-film PV by an average of 7 per cent per year.[9] However, things may be changing more quickly than anticipated. The last few years have seen a dramatic fall in the price of PV, with some silicon modules being sold for as little as $1.85/W at the beginning of 2011, and some thin-film modules on offer for $1.35/W.[10] (Indeed, the low price of some units has led to accusations being made against some Chinese manufacturers that panels are being dumped at below cost price, and this will be an issue to watch for the future.) Should these cost reductions prove sustainable, they will dramatically change the economics of solar, and far faster than anyone was expecting. At these prices, large-scale PV in the Sahara may cost as little as $0.052 / kWh in the next few years.

The trend for CSP is similar. In 2008, McKinsey gave a cost of $0.15–0.16 / kWh for parabolic trough systems,[11] while in 2007 DLR estimated the cost of CSP at somewhere between $0.10 and $0.20 / kWh for installations in North Africa.[12] Because of this wide range of costs, and the difference in their power production profile, it is hard to give a fair comparison of the costs of the various PV and CSP technologies.

Costs for CSP are also expected to continue to fall in the future, thanks to technological change, leading to improvements in efficiency, assembly and manufacture, and the use of cheaper or more durable materials. Economies of scale will play a role in driving down costs too. According to DLR, by 2025 CSP will be the cheapest source of electricity, both in the MENA region (€0.04 / kWh) and in Europe (€0.05 / kWh).[13] McKinsey also believes that costs for CSP will fall – by 3 per cent per year – but still feels that it will be a more expensive technology than PV in the future, expecting that it

Table 4.1 Cost estimates for 100 MW PV installations

	Sahara (Nema, Mauritania)	Sahara (Ouarzazate, Morocco)	Negev	Thar	Sonoran	Great Sandy	Gobi
Module price = $4/W							
Tilt angle = 10°	14.8	18.5	20.6	17.8	18.4	19.1	19.1
Tilt angle = 20°	**14.7**	**17.9**	**20.0**	**17.2**	**17.9**	**18.8**	18.1
Tilt angle = 30°	15.1	17.9	20.4	17.3	18.0	19.4	**17.6**
Tilt angle = 40°	16.1	18.3	21.3	17.9	18.7	20.5	17.6
Module price = $3/W							
Tilt angle = 10°	11.6	14.5	16.6	14.0	14.5	15.5	15.0
Tilt angle = 20°	**11.5**	**14.0**	**16.1**	**13.6**	**14.1**	**15.4**	14.2
Tilt angle = 30°	11.8	14.0	16.5	13.7	14.2	15.9	**13.8**
Tilt angle = 40°	12.7	14.4	17.3	14.3	14.8	16.8	13.8
Module price = $2/W							
Tilt angle = 10°	8.4	10.5	2.6	10.4	10.6	12.0	10.8
Tilt angle = 20°	**8.4**	**10.2**	**12.3**	**10.0**	**10.3**	**11.9**	10.3
Tilt angle = 30°	8.6	10.2	12.7	10.2	10.5	12.4	**10.0**
Tilt angle = 40°	9.3	10.6	13.4	10.7	11.0	13.2	20.0
Module price = $1/W							
Tilt angle = 10°	5.2	6.5	8.6	6.6	6.7	8.5	6.7
Tilt angle = 20°	**5.2**	**6.3**	**8.4**	**6.4**	**6.5**	**8.4**	6.3
Tilt angle = 30°	5.4	6.4	8.8	6.6	6.7	8.8	**6.2**
Tilt angle = 40°	5.8	6.7	9.5	7.1	7.2	9.5	6.2

Source: K. Komoto, M. Ito, P. van der Vleuten, D. Faiman and K. Kurokawa (2007) *Energy from the Desert: Practical Proposals for Very Large Scale Photovoltaics*, Earthscan.

will still cost $0.11–0.12 / kWh by 2020.[14] These figures may be pessimistic, however, with some CSP developers already indicating that they will be able to produce electricity for less than $0.10 / kWh within the next five years.

Despite the optimism, right now, these costs for solar are high compared with those for other renewable energy sources such as wind ($0.04–0.07 / kWh), or combined cycle natural gas ($0.03–0.05 / kWh). Because of this solar power continues to require a measure of government support as it seeks to achieve ever lower prices. This can take the form of capital grants, feed-in tariffs, renewable credits or tax incentives.

Policy environment

As important as the physical resource is the political and economic environment. Germany, for example, is the global market leader in solar PV, despite its high latitude and relatively poor insolation. Another world leader – Spain – while having a reputation for sunshine, still receives less solar radiation than many US east coast states, but until recently was ahead of them in terms of MW of both PV and CSP installed. Consequently, any review of the potential for solar as a large-scale electricity source needs to look at the current and potential financial mechanisms of the key regions.

Local electricity prices matter too. In countries like Italy or states such as California with good sunshine and high residential electricity tariffs (up to $0.19 / kWh in California) rooftop solar will become economic for the private consumer without government support far more quickly that in similarly sunny locations with cheaper electricity. Hawaii for example has the highest retail electricity prices in the United States, largely as a result of the need to import fossil fuels. This makes it a particularly attractive market for renewable energy, and indeed the islands are now seeing a number of new CSP developments.

Subsidies for fossil fuels and nuclear?

The apparently high upfront costs for solar energy ignore certain key features of renewable and fossil energy. Firstly, the cost estimates of fossil fuels do not take account of the financial, environmental and social consequences of the pollution that they produce. For example, the cost of oil extraction does not factor in the potential risk of a catastrophic oil spill, like the recent deep water disaster in the Gulf of Mexico. While the companies responsible will undoubtedly need to pay compensation, it is difficult to imagine that they will be required to bear the full costs of rehabilitating the entire ecosystem, assuming it is even possible. Even if they are, will this be factored into future thinking about the cost of oil? Too often the costs are minimised through lengthy legal proceedings, the use of shell companies, offshore holdings and the like. Following the catastrophic *Exxon Valdez* disaster in 1989, Exxon was finally forced to pay a mere $500 million, plus interest, of the $5 billion punitive damages awarded after the spill (on top of another $500 million in direct pay-outs to affected communities).[15] What is more, legal proceedings are still ongoing and Exxon is still appealing. In the end, it seems a large portion of the costs from these disasters are usually borne by the taxpayer. In Prince William Sound, the site of the *Exxon Valdez* disaster, there are still estimated to be at least 97,000,000 litres of oil in the sand, more than 20 years after the accident.[16] Similarly the consequences of climate change, or geo-political consequences of relying on imported fossil fuels, are not fully incorporated into the price indexes.

Although it is impossible to get definitive figures for how these externalities would affect the economies of conventional power generation (how do you put an accurate cash value on altering the climate, or the devastation of an entire ecosystem or culture?) there have been a number of estimates. Way back in 1998, before the Iraq war and the recent rapid increase in the price of oil, the International Centre for Technology Assessment think tank estimated the true costs of gasoline at $5.6–15.14 per gallon, compared with a price of roughly $1 per gallon at the time.[17] More recently in 2006 Milton Copulos, a former advisor to Ronald Reagan working with the right-wing Heritage Foundation, presented figures to the US Senate showing that the external costs of gasoline were around $5–8 per gallon of fuel or around $829 billion per year.[18] Note that these costs did not even include environmental costs, and were focused on the military and import costs to the US economy. Were the costs of climate change and environmental degradation to be included, the true figure would likely be far higher for the USA alone.

Secondly, aside from these abstract subsidies that result from the failure to cost-in externalities, conventional power continues to receive large amounts of direct state aid. This can take many forms, including tax breaks, money for research and development or price subsidies. According to DLR, in 2007 around 90 per cent of the European Union's research and development budget for energy was still being spent on fossil fuels and nuclear. In many countries, the electricity industry continues to be state run or state-sponsored, with existing power suppliers entrenched and able to take advantage of existing infrastructure. While the renewables industry also receives subsidies, on a global scale these are still small by comparison. The United Nations Environment Programme (UNEP) for example estimates that the fossil fuel industry currently receives an annual global subsidy in the order of $250 billion,[19] with some estimates placing this as high as $500 billion, and one 2007 figure putting this at $49 billion in direct support in the USA alone.[20] Put together this is significantly larger than the entire renewable energy industry, and has been recognised as a barrier by the leading economies with the G20 agreeing in 2009 to work to reduce these levels of public support.[21] According to the European Commission, these additional subsidies for conventional fossil fuels amount to €0.05 / kWh. If this were added to the headline costs, this would place it on a par with CSP.

Finally, and importantly, centralised fossil-fuel and nuclear generation use mature and established technologies, which have benefited hugely from state support, both now and in the past. In many countries they were built up and developed as strategic state-run industries, with the taxpayer supporting the grid, power stations and technological development. Indeed, in many countries the power industry is still state-controlled or run by 'national champions' that enjoy great political influence. Conventional energy is well advanced in its cost reduction curve and changes in the future can be expected to be small. Indeed, costs are likely to increase significantly in the future as the cost of fossil fuels rises. By comparison, solar energy at a commercial scale is a relatively new technology, with a long way to go in its cost reduction process. As I hope is clear from Chapter 2, research is continuing on every front, with a great emphasis on cost reductions and greater efficiencies. New materials, control systems and manufacturing techniques could revolutionise the cost of solar – but it will take time.

Support for the solar energy industry

While the various subsidies and externalities associated with fossil and nuclear fuel are useful to know about (if only to provide an additional justification for supporting the development of the solar energy industry) they are of little short-term comfort to the developer. As has already been said, for the economic situation we are currently in, solar energy remains expensive by comparison with coal and gas (although its costs *are* falling quickly). There is also little realistic prospect of the level of support for fossil fuels being seriously reduced any time soon. Therefore in order to support the solar industry, and enable costs to decline further, support policies for solar will continue to be necessary in the coming years.

This support can take many forms. To begin with some countries are using direct financial measures to promote the development of the solar industry, providing an additional payment for each kWh of electricity produced in recognition of the benefits that solar power will bring. Typical measures include tax relief (popular in the USA), green certificates or carbon trading, and feed-in tariffs (popular in Europe). Each of these has proved successful to some extent at promoting the deployment of solar energy, although each has its own particular pros and cons.

Certificate-based systems / renewable obligations and portfolios

'Green certificates' are so called because they are awarded to the producers of green or low carbon electricity – typically one certificate for each MWh. These can then be sold separately from the electricity to consumers who wish to purchase low-carbon energy, allowing the power producer to make an additional income over and above the value of the electricity. In some cases this may be through a voluntary market, where corporate or domestic consumers are willing to pay a premium for 'green' energy. Increasingly though green certificates form part of regional initiatives and are sold to utilities or companies that have been mandated to source a fixed percentage of their electricity from renewable or alternative energy sources – the basic principle of the Renewable Obligation or Renewable Portfolio Standards popular in the USA and UK. For example a public utility in California might be obliged to provide 33 per cent of its electricity from renewable sources (actually this is the State's 2020 target, but it will suffice as an illustration). It can do this by producing its own renewable energy, or by buying certificates from other renewable energy producers, 'incentivising' the production of low carbon energy. Failure to do either results in a penalty.

Similarly, the UK runs a classic renewable obligation policy. Under this system, utilities are obliged to source a fixed (and gradually increasing) quantity of their electricity from renewable energy. For Great Britain this is set at 11.1 per cent for 2010–2011 (4 per cent in Northern Ireland). They do this by accumulating Renewable Obligation Certificates (ROCs), either from electricity they generate themselves or through those they buy from other generators. Failure to purchase enough ROCs (for example if not enough have been generated) forces the company to pay a buy-out charge, currently set at £36.99/MWh ($57.08/€44.14).[22] The money raised from these buy-out charges is then accumulated and divided amongst all the generators of ROCs, in proportion to how many they produce.[23] So a producer that generated 5 per cent of all the ROCs would receive 5 per cent of all the buy-out penalties. While this scheme was originally intended to be technology neutral, and allow the market to determine the most cost-effective technology, it has since been amended to encourage particular kinds of renewable energy. Offshore wind, solar and marine all now receive multiple ROCs (these are known as banded ROCs), while cheaper forms of low-carbon energy such as landfill gas receive fewer.

To date the ROC-style system seems to have been less successful at spurring renewable energy deployment than the fixed feed-in tariff system, and there are a number of reasons for this. To begin with, until the banding system was introduced, the payback was simply too low to make many emerging technologies economically viable. Secondly, because the price of ROCs can fluctuate with their availability year-on-year this introduces a layer of uncertainty for developers and operators, a problem not encountered with the feed-in tariff which guarantees sale for a fixed price over a fixed period. At the same time though Renewable Obligation systems have their advantages. They are less prescriptive than the feed-in tariff, with less chance of causing unexpected market distortions. Some also argue that they will spur greater innovation in cost reductions over the medium to long term.

Feed-in tariffs

Feed-in tariffs have emerged as one of the most popular support mechanisms for renewable energy in much of the world, particularly for small- and medium-scale applications. Essentially they work by guaranteeing the purchase of renewable electricity at a fixed price, which can be set at a different level for each technology. This fixed rate can gradually be reduced over time, encouraging manufacturers and developers to continuously seek methods of cost reduction.[24] This enables developers to accurately predict the return on their investment and minimises risk. In a new sector such as renewable energy this can be essential in persuading reluctant banks to provide financing.

While feed-in tariffs have come under criticism for stifling competition and innovation by providing fixed prices, and through their ability to create bubbles if certain technologies are over-compensated (see section on Spain below), there can be little doubt as to their success in stimulating the development of renewable energy. Pioneered by Denmark and Germany, they have allowed these countries to dominate the wind and solar markets for many years. It seems highly unlikely that initially expensive technologies such as solar photovoltaics would have developed as far as they have without such stable, technology-specific support, and Germany in particular deserves credit for bankrolling the early development of the photovoltaic and wind industries.

Feed-in tariffs are paid for in two main ways. In most cases, the cost of the subsidy is split between all electricity customers, amounting to a few euros per person per year (most feed-in tariffs are currently in Europe). In other cases, such as in Spain, the costs are not passed directly on to consumers, and are borne directly by the national Treasury. This system of paying the costs directly from the government's budget brings with it a great deal of political risk, as it makes the programme more vulnerable to being cut in difficult times.

Currently more than 50 countries around the world, including 30 in Europe, have introduced some form of feed-in tariff, with more due to adopt the system in the coming years (see Table 4.2). Note that this is an area that is likely to change particularly fast, with many countries moving to slash feed-in tariff levels as the cost of PV falls.

Emissions trading or cap-and-trade

Related to both the green certificate and ROC systems, but more complex, is the concept of emissions trading, also known as cap-and-trade. This is essentially a market mechanism designed to set a cost or price on the emission of carbon dioxide (currently just under €15 a tonne in Europe's Emission Trading Scheme).[25] The first step of the scheme is to set a 'cap'. This theoretically puts a limit on the amount of a pollutant that can be emitted per country or region, with specific quotas then granted to individual industries and then companies and institutions. By enabling the right to emit a certain amount of carbon to be bought and sold, emissions trading allows polluters to 'offset'

Table 4.2 Feed-in tariffs for photovoltaics and CSP in selected countries (2010)

Country	Feed-in tariff (PV) per kWh	Feed-in tariff (CSP)
Germany	€0.29–0.39	
Spain	€0.31–0.34	€0.27
Italy	€0.49	€0.22–0.31
Greece	€0.40–0.50	
France	€0.42	
Portugal	€0.31–0.45	€26.3–27.3 (up to 10 MW)[a]
Czech Republic	€0.49–0.50	
Austria	€0.46	
Turkey	€0.22–0.28 [b]	
United Kingdom	€0.49 (£0.41)	
South Africa		2.10 ZAR / kWh (€0.21)[c]
India	INR 13.70–18.80 (€0.22–0.30)[d]	
Israel	€0.31 for systems up to 50 kW	
South Korea	€0.23–0.265	
Japan	¥50[e] (€0.44)	

a European Renewable Energy Council – Portugal Renewable Energy Policy Review http://www.erec.org/fileadmin/erec_docs/Projcet_Documents/RES2020/PORTUGAL_RES_Policy_Review_09_Final.pdf
b PV Tech 14 May 2010 'Turkey enters the PV arena with €0.28 feed-in tariff' http://international.pv-tech.org/news/turkey_enters_the_pv_arena_with_0.28_feed_in_tariff
c Renewable Energy World 10 April 2009 'South Africa introduces aggressive feed-in tariffs' http://www.renewableenergyworld.com/rea/news/article/2009/04/south-africa-introduces-aggressive-feed-in-tariffs
d Reuters 17 September 2009 'India unveils tariff norms for renewable power' http://www.reuters.com/article/2009/09/17/us-india-renewables-tariff-sb-idUSTRE58G3TW20090917
e Kyodo News 25 February 2009 'Utilities to be premium to solar power providers' http://search.japantimes.co.jp/cgi-bin/nb20090225a4.html

Source: http://www.globalfeedintariffs.com/global-feed-in-tariffs/ (2010)

their emissions by buying credits from other companies or countries which have done something to reduce theirs. For example a cement factory in the UK that has expanded beyond its permitted 'right to pollute' might purchase 500,000 tonnes worth of CO_2 reductions from a steel works in France that has just installed more energy efficient equipment or a renewable energy source, providing an extra incentive for energy efficiency, and allowing the cement factory to continue as normal. Similarly if a steel mill's production falls, it may have excess credits which it can then sell to generate extra income. The theory goes that it does not matter where in the world the emissions are prevented, and that this finds the lowest cost way to reduce greenhouse gas emissions.

There are currently two main emissions trading systems in operation: the European Emission Trading Scheme (ETS – which is a cap-and-trade system in Europe) and the Clean Development Mechanism (CDM) and Joint Initiative (JI) mandated under the UNFCCC's Kyoto Protocol. Of these, the most interesting for large-scale solar energy in the desert is the CDM. The CDM is a mechanism whereby emerging and developing countries (Annex II of the Kyoto Protocol) can generate emissions reductions credits from energy efficiency and renewable energy projects and sell them to developed Annex I countries that need to reduce their emissions. Put simply, those projects which think they may be eligible must register with the CDM programme and have a review conducted. This review then determines how much carbon emissions will be reduced by, and certifies that the project is suitable. Once this has happened the project may be accredited to the CDM system and can accrue and sell CDM credits. This could provide a financial mechanism to support the deployment of renewable energy in developing countries without them having to implement expensive feed-in tariffs or internal renewable obligation systems.

Although the Emissions Trading Scheme is confined to Europe, emissions reductions produced under the CDM (so-called Certified Emissions Reductions or CERs) can often be traded into the ETS.

Just as this book was going to press, a second emissions trading or cap-and-trade system was formally established in California, and moves were under way to link the European ETA and the California cap-and-trade, presumably allowing each to buy and sell credits generated in the other.

Box 4.1 Problems with carbon trading

While at first glance the concept of swapping emissions reductions in one location for increases in others may seem fine, in practice the concept has come in for serious criticism from many sides. It is beyond the scope of this book to provide a detailed criticism of carbon trading but there are a number of serious concerns that will be briefly mentioned. It is important to raise these issues here, in part because they are interesting, but also because it is possible that these problems will destabilise the various carbon trading schemes in the coming years.

Firstly, while a carbon-trading system may work within a strictly controlled region or industrial sector, how can it be controlled on a global basis? In theory at least a closed system in a single state or country provides a cap on emissions within that bloc, and trading happens beneath this cap. This means that tight control can be kept on overall emissions, and it should be possible to ensure that all the trading takes place within an ever-decreasing pool of emissions, leading to efficient reduction in pollution. This is simply not the case with large international emissions trading schemes. With the international CDM market there is no cap, only trade, since there is currently no obligation on developing countries to reduce their emissions. This means that all emissions reductions under the CDM are theoretical, measured against a possible baseline scenario, i.e. how much more *might* have been emitted without this project. This makes it extremely hard to quantify how successful it is.

Similarly, since many of the projects involve things such as energy efficiency or new power stations for growing economies, there is no guarantee that the projects earning the credits would not have gone ahead anyway. This added value is known as 'additionality' and although it is theoretically incumbent on any new project to demonstrate this additionality, it is very hard to see how it can be done accurately. For example, a recent CDM story has centred on the 1320 MW Mundari coal plants, part of a massive coal-power generation complex being developed by the Reliance Power company in Gujarat, India (part of Adani Group). In this case the company has proposed that the new power station they are building will be an efficient super-critical facility, which will release less CO_2 per unit of energy than a sub-critical coal power station.[26] Because the Adani company *could theoretically have* installed a less efficient power station, they have been approved as a CDM project and can now sell the emissions 'reductions' to companies in the rich world, meaning that those rich-country companies will not need to reduce their emissions either. Of course, no real emissions reductions have actually occurred. Indeed the net amount of carbon being released into the atmosphere has actually increased, just perhaps not as much as it might have. There is also every reason to believe that Adani might have built a super-critical power station without the CDM money, since they are more efficient and many new coal power stations are super-critical. It is also hard to imagine a project of this scale getting financed on the basis of a possible CDM payout. So in effect, the CDM has provided a direct subsidy to developing a coal power station. In such a circumstance it seems a poor substitute for concerted action and technological innovation in developed countries. In fact it could be argued that it is actually harmful, since the benefit of installing a more efficient coal power station has been counteracted by inaction elsewhere. Furthermore it has provided a direct subsidy for coal power, at a time when we know we need to phase it out. Sadly, there are many such examples, some of which seem truly ridiculous. Recent documents released on the website Wikileaks have shown that in conversations between the US Embassy in India and senior Indian business people it was generally assumed that not one of the projects in India which was granted CDM credits would not have gone ahead anyway.[27]

Next there is the question of whether it is fair or wise for rich countries to delay their transition to a low carbon economy by paying poor countries. Industry in the EU often voices concerns that greater emissions limits and environmental standards may lead to so-called carbon leakage, where heavy industry merely moves overseas where they can pollute all they like. This is a fair concern. However, many of these same companies are in favour of carbon trading. This seems to be a huge contradiction. If heavy industry in Europe can only continue to pollute by buying credits that fund modernising companies in

Continued ...

Box 4.1 continued

Asia and elsewhere to install the most efficient equipment available, are they not simply providing a direct subsidy to their competitors, avoiding short-term change themselves while paying for the modernisation of industry elsewhere?

Thirdly, there is tremendous room for corruption and dodgy accounting. CDM and other similar credits must be approved and checked by third party assessors. It is not hard to imagine assessors being encouraged to approve or exaggerate emissions reductions when hundreds of millions of dollars are at stake. As with other systems, these will only be as accurate and transparent as those conducting them, and already stories are beginning to filter out that organised crime has spotted a potentially lucrative source of income. Should this come to pass it could further undermine confidence in the system, with consequences for renewable energy developers. Thus it will be important for potential investors to be aware of potential problems with the CDM.

'Gaming' of the carbon trading systems is also a serious concern. One of the principle sources of CDM accredited projects to date has been the destruction of the industrial gas HFC-23. HFC-23 is a powerful greenhouse gas produced as a by-product of making the refrigerant chemical HCFC-22. Capturing and destroying the gas at the point of manufacture prevents it being released into the atmosphere and contributing to climate change. So far so good. The problem is that the money paid for the destruction of the gas (€15 per tonne of CO_2-equivalent) is so high, that destruction of this gas becomes more profitable than producing the cooling chemicals. There is strong evidence that manufacturers have seriously overproduced HFC-23 in order to earn subsidies, while at the same time selling their destruction as an offset to the European ETS, allowing continued pollution from European countries. Thus pollution is increased and expensive payments are made for a worthless service. So serious has the situation become that the European Commission finally voted to ban HFC credits from the ETS, but not until 2013.[28, 29]

Finally, and crucially, even within an apparently closed system like the ETS, emissions quotas (rights to pollute) were often originally handed out to industry for free. Because there was no initial cost to acquiring these rights, major polluters and nations simply submitted inflated estimates of how much pollution they would generate. This produced a glut of emissions credits, in the first round of the ETS, resulting in a low carbon price. Furthermore since these were handed free to major polluters, but still have a value, it is essentially a subsidy for the most polluting industries. It is also a penalty against newcomers, since they must purchase credits to pollute from old industries that received them as a hand-out. While some of these problems are being addressed by auctioning a proportion of the credits in the second and third rounds of the ETS, the fact that polluters can carry over excess permits from the first round, coupled with a dramatic fall in emissions as a result of the financial crisis and lower industrial activity, means that the ETS will fail to adequately price the value of carbon emissions for many years, and continue to provide support to polluting industries – exactly the opposite of its stated aim.

Other economic issues

Since solar energy is in direct competition with fossil fuels, it is not just the cost of the renewable installation that matters but the cost of the competition. Solar fuel is free, but fossil fuels are not. As oil, coal and gas prices rise, the economics of alternatives can look increasingly favourable. While the price of these key fossil fuels has recently fallen thanks to reduced demand caused by the global financial crisis, the long-term trend is almost certainly up, and already at the time of writing the price of oil was again approaching $80 a barrel. This is already very high compared with predictions of just a few years ago. A report published by the IEA in 2005 concluded that the long-term price of oil would be between $35 and $45 per barrel. Similarly in its Trans-CSP report, published in 2005, DLR estimated a conservative baseline price of $25 per barrel for oil, and $49 per tonne of coal. Both estimates now look highly optimistic. Instability in the Middle East may have pushed oil prices over $125 / bbl, but even before the recent uprisings in North Africa oil was nearing $100 / bbl. At the same time the price of coal has soared on the back of demand from Asia and floods in the coal-producing regions of Australia. In fact the price of coal recently set a new record in April 2011 when a deal was struck for $130 per tonne between mining group Xstrata and a Japanese company.[30] If fossil fuel prices remain high, this increases energy costs for consumers, and reduces the relative cost of solar energy, providing an incentive for investment, both in generation capacity and in research and development.

Status of regional markets and support policies

Countries all over the world are beginning to recognise the benefits that a move to solar energy will bring, and the last few years have seen a wide number of support measures being implemented. This section looks at some of the key national markets region by region, and examines the policies they have put in place to encourage development.

Europe

Currently Europe is the world leader in solar, both in terms of total CSP and PV deployed, and the annual rate of installation. While they are not strictly in the desert regions, European countries and corporations are playing such a crucial role in the development and deployment of large-scale solar energy that it would seem perverse not to include them, and in particular to look at the policies which have managed to spur growth.

Germany

In PV, Germany is by far the market leader, thanks to its feed-in law which established a fixed payment of €0.53 for every kWh of solar electricity produced (this law was first established in 1990, updated in 2000 and has been amended a number of times since).[31] From 2005 to 2009, Germany installed roughly 9000 MW of PV, and installations continue at a rate of over 3000 MW per year.[32] In fact so successful has this scheme been that the German government has reduced the feed-in tariffs twice to reduce costs, and recently reduced them again by around 15 per cent in 2010. Prior to this latest reduction they stood at €0.32–0.43 / kWh depending on the size of the system.

As well as the rooftop and domestic market, Germany is home to some of the largest ground-mounted PV plants yet constructed, including two over 50 MW at Strasskirchen and Lieberose. Despite this, the country is unlikely to be a major market for large-scale centralised solar in the longer term future, thanks to its high latitude, relatively poor insolation and high population density. In particular it is outside of the 'sunbelt' and unsuitable for commercial CSP or concentrating photovoltaic developments, which require constant direct sunshine (although the country has opened a 1.5 MW power tower (CSP) plant as a test and demonstration facility).

German companies, however, are likely to remain at the forefront of both CSP and PV. In CSP particularly many of the most important manufacturers of equipment and components such as Schott Solar, Flabeg and Slaich Bergmann are German. This is reflected in initiatives such as DESERTEC (see Chapter 6 on upcoming projects), which has a large number of German partners. In photovoltaics too, Germany is host to a range of international manufacturers along with home-grown companies such as Q-Cells, SolarWorld and Conergy. The crucial factor for German PV manufacturers will be if they can survive in the face of competition from China and Taiwan, as the global PV industry enters a period of scaling up and consolidation.

Spain

Of all the countries in Europe, Spain looks to have the biggest long-term potential for large-scale solar. The country is already a world leader in the field, with numerous multi-megawatt PV and CSP installations either operational or under development.

For PV, the situation was driven by a 2004 Royal Decree which established a new feed-in tariff for photovoltaics, set at €0.44 / kWh. In 2007, the upper limit on the size of the installations was changed from 100 kW to 10 MW. This was generous, and in hindsight too generous. Installations rocketed far beyond the government target, payments to the PV industry increased to €2.5 billion for a year, and even poorly designed systems could make a profit.[33] This, coupled with a severe recession and the fact that the costs were borne by the Treasury, rather than by electricity consumers, prompted the Spanish government to rush through yet another decree in 2008, reducing payments to €0.32 / kWh, and introducing a pre-registration system to ensure that caps were not exceeded. The result was a crash of over 95 per cent in the PV market. From seeing over 3000 MW of new PV installed in 2008, this fell to just around 100 MW in 2009,[34] and columnists and free-marketeers around the world proclaimed the failure of the feed-in tariff. Indeed, for those who fear that feed-in tariffs cause an unhealthy distortion in the market, Spain is exhibit 'A'. These fears, however, may be overdone. The Spanish market is now recovering, with 450 MW of installations in 2011. Importantly though, it seems that the days of large ground-mounted PV systems may be over in Spain, and the focus has turned firmly to domestic and rooftop generation.[35] This is not necessarily a bad thing, however. In a country such as Spain, ground mounted systems have significant consequences for land-use, wildlife and agriculture. It may be that rooftop solar is a more sensible long-term use. One of the key achievements of the feed-in tariff in Spain though, despite its boom and bust, has been to develop an industry. Spanish companies are now exporting their equipment and experience around the world, while within Spain itself, PV has become 'normalised', and accounts for around 3 per cent of national electricity requirements.

On concentrating solar, things are quite different. As with photovoltaics, the renaissance in this industry began in 2004, when the government set a target of 500 MW of CSP by 2010, and introduced the feed-in tariff for solar electricity. For CSP plants up to 50 MW this is set at 26.9 eurocents / kWh, falling to 21.5 eurocents after 25 years. This feed-in tariff is increased yearly with inflation, minus one percentage point.[36] In addition, the feed-in tariff allows natural gas back-up of 12–15 per cent to be used.

While this incentive has existed in some form since 2004, it has taken several years for projects to come to fruition. However, the stage is now set for rapid growth, and nearly 2000 MW of new CSP is under construction in Spain.

In fact the sudden success of the Spanish CSP industry could bring with it some complications. Currently, the National Commission of Energy keeps track of the register of installations. When 85 per cent of the government target is reached, the Commission must decide on how much longer new projects can claim the highest level of feed-in tariff. This is causing a rush to complete projects and is generating uncertainty in the industry. To avoid this, some industry advocates have suggested that the target be changed to 1000 MW of CSP per year, which will help the industry maintain a steady pace. It remains to be seen exactly what will happen, but the Spanish government is no doubt keen to avoid a repeat of the solar PV situation outlined above, where changes in the law caused a boom in 2008, followed by a sudden bust as feed-in tariffs were reduced and new caps were placed on growth.

Italy

Despite having an excellent solar resource in the south of the country, Italy has been relatively slow to move into the large-scale solar arena, particularly in terms of CSP. In early 2001 the Italian government allocated €100 million to the National Agency for New Technologies, Energy and the Environment (ENEA) for research into new energy. This was later reduced to €50 million with the money spent on developing a molten salt storage system for parabolic trough technology, and on research into hydrogen production by thermo-chemical means. Following this, in 2004, ENEA and Italian utility ENEL signed an agreement to construct the 5 MW Archimede CSP plant in Sicily.[37] This is a parabolic trough plant, and, when completed, it will be Italy's first.

In 2008, Italy introduced a feed-in tariff for CSP set at €0.22–0.28 / kWh, depending on the proportion of a plant's energy which is derived from CSP.[38] It remains to be seen if this will be at a sufficient level to spur significant growth.

The PV market is rather more advanced. Following the introduction of a feed-in tariff of up to €0.49 / kWh for solar PV, photovoltaic installations in Italy have soared, including at least one installation of more than 20 MW. In 2009, 770 MW was installed,[39] a growth of 114 per cent on the previous year. Such has been the growth in the industry, and the general fall in the cost of solar, that in 2010 the government announced it would slash the level of feed-in payments by 20–30 per cent in 2011, and a further 6 per cent in 2012 and 2013.[40]

Other countries in Europe

Other countries in Southern Europe, particularly Greece, Portugal and the south of France are also seeking to encourage large-scale solar development. All have existing feed-in tariffs for solar electricity, and Portugal in particular has several large-scale PV installations, including the 62 MW Moura plant and the 11 MW Serpa plant. The Czech Republic too has emerged as a significant market thanks to its new feed-in tariff, enjoying over 700 per cent growth in installations from 2008–2009.

Table 4.3 shows the combined solar capacity and annual installation rate for PV in key countries in MW (according to European Photovoltaic Industry Association and SolarBuzz).

United States of America

Unlike Europe, the USA has not favoured the use of feed-in tariffs to encourage renewable energy, instead preferring mechanisms which set the target in a technology neutral way (this may be about to change, at the time of writing there were suggestions that the Obama administration was reviewing options for feed-in tariffs, although the odds of it successfully introducing a national one seem unlikely). Nationwide, support measures for renewable energy include significant investment in research and development, grants or loans to specific projects as well as tax relief on renewable installations. This latter comes in two forms known as the Investment Tax Credit (ITC) and the

Table 4.3 PV installation rates in key markets 2007–2010 (in MW)

Country	2007	2008	2009	2010 estimates from SolarBuzz
Germany	1271	1809	3806	6600
Spain	560	2605	69	103
Japan	210	230	484	620
USA	207	342	477	970
Italy	70	388	711	1300
Czech Republic	3	51	411	400
China	20	45	160	480
India	20	40	30	n/a

Source: 2007–2009 figures, European Photovoltaic Industry Association, 2010 figures from SolarBuzz

federal Production Tax Credit (PTC). The Investment Tax Credit provides tax relief on capital expenditure on renewables, while the Production Tax Credit provides tax relief on the sale of electricity from renewable sources (per kWh). Crucially there is also an option to take the ITC as an upfront payment of up to 30 per cent of the cost of the installed system.[41] There has been a great deal of uncertainty about this support mechanism as it has traditionally only been extended for periods of two years at a time, introducing an element of uncertainty to the market. In 2008, however, it was extended through to 2017, providing some degree of predictability for investors.

Other key drivers have been state level policies, particularly Renewable Portfolio Standards (RPS) that require investor-owned utilities to source a fixed percentage of their electricity from renewable sources by a given date. Altogether more than 30 states and Washington DC have an RPS, but the ones of most interest to large-scale solar are California (20 per cent by 2017), Nevada (15 per cent by 2013), New Mexico (10 per cent by 2011) and Arizona (15 per cent by 2025).

It has been suggested by several commentators that neither the Investment Tax Credit, nor the Renewable Portfolio Standards can provide sufficient support to the renewables industry by themselves, and in those times that the ITC has been about to lapse investment has dropped off, lending support to this view.

To date, despite some debate, there is no current likelihood of the United States introducing a national RPS or cap-and-trade system. In late 2010, however, California introduced a cap-and-trade system that once up and running should be the second largest in the world, after the European Emissions Trading Scheme (ETS). First mooted as part of Governor Schwarzenegger's Global Warming Solutions Act in 2006, the Bill sets a binding target for reducing greenhouse gas emissions to 1990 levels by 2020, meaning a cut of 25 per cent over 2008 levels and indicates that market-based mechanisms should be used to help reduce emissions. Starting in 2012, the scheme hopes to cover around 85 per cent of the State's emissions by 2015, and will also cover out-of-state electricity suppliers.[42] Given the cultural and economic importance of California, both to the United States and the world as a whole, this cap-and-trade programme is likely to be one of the most heavily scrutinised in the world and will likely be the focus of a great deal of debate in the USA. Given the strength of negative feeling amongst some sections of US politics to environmental regulation, we can only hope that the discussion centres on the facts rather than on ideology. Table 4.4 summarises how the RPS is being implemented by each US state.

Leaving aside political support policies, what is actually happening in the US? In 2009, the United States installed 485 MW of photovoltaics[43] and at least one small CSP installation. The sector is expanding rapidly though, with one analyst estimating that there were nearly 1000 MW installed in 2010 and predictions from the European Photovoltaic Industry Association suggest that the USA could install up to 2000 MW of PV a year by 2013, including both large-scale and decentralised installations (if the policy environment is right, this could be up to 4500 MW).[44]

On the CSP front, there are currently around 9.5 GW of developments in the pipeline, several of which have been supported with large capital grants from the Federal Government. CSP is an important area for the USA and it is rapidly emerging as a world leader in the field, along with Spain. For more information see Chapter 5 on existing and upcoming projects.

China

Although China has tremendous potential (and need) for large-scale solar, to date its policy reaction has been minimal. In this respect China remains the great unknown, but the consensus seems to be that it will emerge as a major market for solar in the coming years.

Certainly in the manufacture and production of PV, China is already a dominant force. In 2009, Chinese companies produced around 4000 MW of silicon solar modules,[45] nearly 40 per cent of global PV production.[46] In 2010, China is expected to manufacture around 7000–8000 MW of PV modules.[47] The vast majority of this is exported, and domestic demand is muted. Partially as a result of slowing overseas demand, in 2009 the government realised the benefit of spurring local growth. The first step by the government was to introduce an upfront payment of 15–20 yuan ($2.2–3) per watt of building integrated photovoltaics installed. As a result 91 MW of PV were approved by the government in the first six months. While this is certainly growth, for a country the size of China it is still a poor uptake, so following this, in July 2009, the government introduced its Golden Sun Programme in which the government would subsidise 600 MW of PV installations over the next 2–3 years (up to 2012). The government would pay 50–70 per cent of the costs.

Table 4.4 Renewable Portfolio Standard by US state

State	Amount	Year
Arizona	15%	2025
California	33%	2020
Colorado	30% (at least 3% from solar)	2020
Connecticut	27%	2020
Delaware	20%	2019
District of Columbia	20.4%	2020
Florida	7.5% (Governor's Bill mandates 20%)	2015
Hawaii	40%	2030
Illinois	25%	2025
Iowa	105 MW	Passed in 1983
Kansas	20%	2020
Maine	10%	2017
Maryland	20% (at least 2% from solar)	2022
Massachusetts	15%	2020
Michigan	10%	2015
Minnesota	25%	2025
Missouri	15%	2021
Montana	15%	2015
Nevada	25% (including 6% from solar)	2015
New Hampshire	23.8%	2025
New Jersey	22.5%	2021
New Mexico	20%	2020
New York	25%	2013
North Carolina	12.5%	2021
North Dakota	10%	2015
Ohio	12.5% from renewables	2025
Oregon	25%	2025
Pennsylvania	18%	2020
Rhode Island	16%	2019
South Dakota	10%	2015
Texas	5,880 MW (5% of projected demand)	2015
Utah	20%	2025
Vermont	25%	2025
Virginia	12%	2022
West Virginia	25%	2025
Wisconsin	10%	2015

Source: http://www.pewclimate.org/sites/default/modules/usmap/pdf.php?file=5907

As a result a total of 643 MW of projects were approved, of which 35 are utility-scale projects totalling 296 MW.[48] Crucially and controversially, these projects were approved in part on the basis of the manufacturer, with some commentators suggesting that domestic Chinese producers have been given preferential treatment.[49]

The National Development and Reform Council has also undertaken research into the possibility of introducing a feed-in tariff for solar (PV and CSP), with a range of figures up to 4 RMB / kWh for demonstration projects expected to apply if and when it is finally introduced.[50,51] So far, however, this feed-in tariff remains firmly theoretical, and the government warns it could be some years before it is introduced.

Looking to the future, the plans quickly start to add up. According to some reports, the new energy stimulus plan being developed by the government calls for a total of 20 GW of PV by 2020.[52] How many of these will be realised is anyone's guess, but judging by the explosive growth of the Chinese wind industry they do not seem unrealistic.

Importantly for the rest of the world, China may not simply be a source of solar panels; it may also provide a market. At present China has just a few hundred MW of installed PV capacity, including six power stations greater than 10 MW and one larger than 20 MW. In 2009, US company First Solar announced a deal to begin work on what could eventually be one of the largest PV installations in the world. The plant, at Ordos City in Inner Mongolia, some 600 km west of Beijing, would be built in stages. The first stage will be relatively modest at just 30 MW, but the next four stages could bring the total capacity up to 2000 MW by 2019.[53] Whether or not this project ever sees the light of day is still unclear, however, as local Chinese producers have voiced opposition to a foreign company being granted such a large deal. However, at the time of writing, First Solar has signed a Memorandum of Understanding with China Guangdong Nuclear Solar Energy Development Co., suggesting things may be looking up for the project.[54]

With CSP technology, things are not quite so advanced. In its 11th five-year plan, the Chinese National Development and Reform Commission (NDRC) called for 200 MW of CSP installations in Inner Mongolia, Xinjiang and Tibet by 2010, with a 25-year power purchase agreement offered. It is unclear if there has been any movement towards developing these projects, and it is unlikely they have all been completed. Aside from the demonstration projects there are no definite CSP projects planned in China – just a couple of small experimental systems.[55] It seems reasonable though to assume that in the event of a properly regulated feed-in tariff or other policy incentive, China would emerge as a major market in the field of CSP too.

Part of the difficulty when writing about China is that although the scale of the plans may be highly ambitious, and the technology and resources are certainly there to make them happen, it is, in my view, difficult to separate the hype from the reality. As with anywhere in the world, many high-profile projects may not come to fruition, but the lack of transparency in China, combined with poor reporting makes the situation more complicated than usual. The Dongtan eco-city is a good example. If I were writing in 2006 it would likely take pride of place as one of the great visions for a sustainable future. The huge, low-carbon city designed by Arup and located on an island just off Shanghai was supposed to be an example to the world and just the first of many. Sadly nothing has yet happened. Recent reports suggest that the land has been cleared but that there is no indication of progress. The project is on hold, perhaps for good, and even its designers have reportedly not been told why. The project office has reportedly been closed down.[56]

This is not to say that great and impressive things in the field of renewable energy are not happening in China, or that work on Dongtan will not restart, simply that the media frequently talk of breath-taking plans and world-beating achievements, which may or may not actually happen.

India

In India's 2008 National Action Plan on Climate Change, initial plans were set out for the deployment of large-scale solar. Research and development to achieve cost reductions and a solar energy research centre have been proposed, as has a target of 4000 MW of solar by 2017, including a number of CSP pilot projects. Following this, in 2009, India unveiled the Jawaharlal Nehru Solar Mission, a national project with the aim of making PV and base-load CSP competitive with fossil fuels by 2022, and installing up 20 GW of solar by the same date (for more information on the details of these proposals, see Chapter 6).[57]

In the meantime, in late 2009, the Indian government established a national feed-in tariff system, designed to provide a return on investment of 19 per cent before tax. This has been estimated at INR 13–18.8 / kWh for solar technologies, with an (annual) cap of 10 MW in each state.[58] This is, it has to be said, a rather disappointing cap for a country the size of India, with its growing demand for energy and the solar resources available to it. In 2009, India installed just 30 MW of PV and no new CSP.[59] Nevertheless there is increasing interest in the potential of India for large-scale solar, and as an Annex II country under the Kyoto Protocol, Indian projects will also be able to seek financial support under the Clean Development Mechanism. Individual states too may seek to introduce their own systems with more ambitious caps, while the National Solar Mission mentioned above may introduce further changes to the national feed-in tariff policy.

In manufacturing, India has recently emerged as an important producer of solar cells, shipping around 700 MW in 2009.[60] An important component of the solar action plan involves supporting and growing this industry not only to support India's ambitions, but also to maintain exports to the rest of the world.

The Middle East and North Africa (MENA)

The countries of the Middle East and North Africa have some of the greatest potentials for large-scale solar energy deployment. With purchase-power adjusted incomes ranging from $145,000 per capita in Qatar to less than $5000 per capita in Morocco, they also encompass the complete range of economic situations.[61] As elsewhere in the world, in addition to instigating their own measures, the poorer countries may be eligible for support under the Clean Development Mechanism, while in the richer ones, such as the UAE, large-scale solar projects are already being constructed as demonstrations of what these countries can achieve. Currently the Middle East and North Africa are undergoing a period of intense political turmoil and it remains to be seen what impact this might have on the development of solar energy in the region. So far the two countries most advanced in this area – Morocco and Algeria – have been relatively calm, although this could all change. Far from seeing the Arab Spring as a negative for solar power, it is equally likely that if more stable and democratic governments can be established this could provide a boost for their economies and renewable energy in general.

Algeria

Along with Morocco, Algeria has perhaps the most comprehensive strategy for developing solar and renewable energy in North Africa, with a national goal of 10 per cent of all energy from renewables by 2025 and a 5 per cent target by 2015. In 2004, the country also became the first non-OECD nation to implement a feed-in tariff for CSP. Developers of ISCCS plants could earn a premium depending on the solar component of their plant. Facilities with 5–10 per cent solar input would receive an additional 100 per cent of the reference price, while projects with over 20 per cent would receive 200 per cent. Finally, to help spur this development, and also to increase co-operation with European partners who are interested in developing large solar installations, the Algerian government formed a special company called New Energy Algeria (NEAL).[62] As a result of these activities one CSP plant is currently under construction and at least two more are planned by 2015.

In photovoltaics, things are also beginning to move. In late 2009, the national energy company Sonelgaz put out a call for a photovoltaic cell manufacturing plant to be sited in Algeria, with an output of around 50 MW. Reports have suggested that this could be completed around 2012 and would be the first solar cell manufacturing plant in North Africa.[63, 64]

Morocco

When I wrote an article on this topic not long ago, I said that Morocco had no national support programme for large-scale solar electricity. Just a couple

Figure 4.1 Small-scale photovoltaic installation in Morocco (source: Isofoton).

of years later, it is emerging as one of the brightest prospects for large-scale solar in the region. In early 2010, it announced that it had reached a $9 billion deal with the World Bank, European Commission and German government to develop solar technology, with the aim of providing 38 per cent of its electricity from solar by 2020 (Figure 4.1).

Plans are under way for an initial 2000 MW of concentrating PV spread across five sites with a combined area of 10,000 hectares. According to press reports, the plants will be located near the towns of Ouarzazate, Ain Bin Mathar, Foum Al Oued, Boujdour and Sebkhat Tah, and could form one of the first physical phases of the DESERTEC Initiative (see Chapter 6).[65]

There are also other ambitious plans for photovoltaic deployment. In 2009, in a deal between Morocco and Spain, the Spanish company Isofoton agreed to develop over 1200 separate PV installations in the Errachidia and Ben Guerir regions. Since each of these will have a capacity of 0.5–1 MW this will represent over 1 GW of new photovoltaics.[66] The plan will also see PV systems installed on the offices of the Office National d'Electricite (ONE).

Currently Morocco has just one CSP plant under construction, again supported by a $50 million loan from the World Bank. This is a 470 MW ISCCS plant (with a 30 MW solar component) at Beni Mathar that was completed in 2011. Work on the project began in 2004 when a tender was released to the industry. This was won by Abener (a subsidiary of Abengoa) and work is under way. By the time this book is published it should be fully operational.[67]

Egypt

The most populous of the Arab and North African countries, Egypt is also one of the poorest. As a result it has a great need for energy in order to help develop its economy. As far back as 1995, feasibility studies on parabolic trough and power tower technology were conducted and in 1996 it was decided that the country would develop a 140 MW system, with a 20 MW parabolic trough component (see the Kuraymat plant in Chapter 5). Following funding from the World Bank, contracts for the project were finally awarded in 2007 and construction is almost complete.[68] In June 2010, it was announced that the Egyptian government intended to build an additional 100 MW CSP station, near Kom Obo in the south of the country.[69] As yet, however, Egypt has no national support policies to encourage the further development of solar, and it will be fascinating to see if the new Government makes renewable energy development a priority.

United Arab Emirates

Currently under construction, the Masdar eco-city near Abu Dhabi is one of the most ambitious environmental projects under way anywhere in the world. Billed as an attempt to construct an entirely new and sustainable city in the desert, the project has enjoyed the support of the royal family. Thus while there are no specific support measures in the Emirates as such, the technology does enjoy considerable patronage. As well as sustainable architecture and decentralised PV, the city will have a dedicated solar supply. The first 10 MW PV plant was completed in 2009, and plans are under way for a further 100 MW CSP plant.[70] Rather unbelievably, it has been reported that both the CSP and PV plants will be eligible for credits under the Clean Development Mechanism.[71] Given that the United Arab Emirates is one of the richest and most polluting countries in the world, this seems incongruous to say the least. For more information on Masdar see Chapter 6.

Israel

With a thriving high-tech centre, endless sunshine and a need to reduce its energy imports, Israel is a natural candidate for large-scale solar. Despite being a small country, there is more than enough desert available in the Negev to provide for its future. In 2006, Israel introduced a feed-in tariff for independent power producers generating solar electricity. This is set at $0.204 / kWh for projects ranging from 100 kW to 120 MW and $0.163 for projects over 20 MW. This is fixed for twenty years, and in the projects over 20 MW, allows for up to 30 per cent back-up generation from natural gas. In

2010, this was supplemented with a feed-in tariff of $0.42 (€0.31)/kWh for small residential systems.

In 2007, the Israeli government augmented this by issuing a call for proposals to build 220 MW of CSP and 15–30 MW of photovoltaics in the Ashalim region.[72] The projects were put out to tender in 2010 with construction to be completed by 2014.[73]

Alongside the government support, Israel also has a strong tradition of self-sufficiency in its kibbutzim, and several of these are providing a ready market for small- and medium-scale off-grid solar initiatives. CSP and PV projects are currently under way at a number of kibbutzim in the Negev Desert, including the Kitura, Yavne and Samar settlements.

Israeli businesses and companies have also played an important role in the development of the solar power industry, through pioneers such as Luz Engineering, Solel and Brightsource Energy.

South Africa

In 2009, the South Africa National Energy Regulator approved a set of feed-in tariffs for renewable energy. Called REFIT, the scheme provides 2.1 R / kWh for CSP, fixed for 20 years. The scheme will be reviewed in 2014, and every three years after that.

Alongside the feed-in tariff the South Africa government has set a target of 10,000 GWh of renewable electricity by 2013. At current consumption, this would supply roughly 5 per cent of South Africa's electricity requirements.[74]

In 2010, the South African government announced exciting plans for a 5000 MW series of solar systems in the Northern Cape Province. With an initial target of 1000 MW of CSP and PV systems envisaged by 2012, the whole array is hoped to be operational by 2020. Total investment in the project could reach up to 200 billion rand (€20.4 billion).[75]

Australia

Disappointingly, Australia has for a long time been the country with the greatest solar resource but with the least to show for it. With a highly developed and expanding economy and abundant sunshine, Australia should be at the forefront of the renewable revolution. So far this has not been the case, and the combination of high energy consumption with a reliance on coal and natural gas has given Australia one of the highest per capita carbon emissions in the world. At the end of 2009, Australia had installed 115 MW of PV and 1 MW of CSP, meaning that solar provided just 0.1–0.2 per cent of total electricity production.

As has been mentioned, it is not a lack of sun or space that has held back development in Australia (deserts cover some 85 per cent of the nation), but the economic and political environment. Up until 2009, support for solar power was provided through a system of rebates of up to AU$8000 for installing solar panels on homes and community buildings. Rebates of up to AU$50,000 were available for schools. While some growth was encouraged, it has been disappointing and the rebate system has now given way to a green certificate system, part of the Mandatory Renewable Energy Target (MRET).

The MRET, which was introduced in 2010, is a federal policy that aims to see Australia generate 20 per cent of its electricity from renewable sources by 2020 (45 TWh). Like the systems in the USA and the UK, the MRET requires wholesale purchasers of electricity to purchase Renewable Energy Certificates, providing an incentive for renewable energy production. For small-scale solar power, multiple RECs will be provided, up to five per MWh for the first 1.5 kW of installed capacity. Judging by the mixed success of incentives such as this in other countries, it remains to be seen if a green certificate system will be effective in encouraging solar power in Australia, and critics have been quick to point out that the federal government continues to lack ambition in this area. This is most easily seen in the CSP front, where arguably Australia has even more to gain. Here, the government has budgeted AU$1.3 billion in support over the next 6 years for four solar plants with a total capacity of 1000 MW. Given the large subsidies to the fossil fuel sector (which according to one study in 2003 were estimated at around AU$8 billion for the total fossil fuel industry),[76] this does not seem a vast amount.

However, Australia, like many other countries, is a collection of states, and many of these are now introducing their own measures to support solar power. While a federal feed-in tariff has been put before the Australian Senate (where it still lies), states have been moving ahead with their own FITs. Two types of FITs are in operation in Australia, those that pay a premium on all electricity produced whether used by the generator or exported (gross), and those that pay a premium only on exported electricity (net). Table 4.5 shows details of the various feed-in tariffs that have been enacted or are in consideration in Australia.

As a result of these efforts, domestic solar installation can be expected to increase significantly in the coming years, although it remains to be seen if this level of support can cause them to really reach their potential. There are also a number of large-scale projects in the pipeline and more of these will no doubt come on-line as the MRET begins to take effect. One of the first and most interesting large-scale solar projects will be a 154 MW concentrating PV (CPV) plant in Victoria. The plant will use 500 times magnification and should be able to generate sufficient electricity for 45,000 Australian homes (150,000 Japanese homes!).

CSP may be about to receive a boost too, with AREVA's purchase of Australian company Ausra, and its announcement that it will be developing a 44 MW linear Fresnel reflector system.[77] This is just one of several solar-fossil fuel hybrid plants Areva is planning in Australia, using either coal or natural gas

Table 4.5 Feed-in tariffs in Australia (updated in 2012)

State	Gross or Net	Max size allowed	Rate of FIT per kWh	Duration of FIT	Current status
Victoria	Net	5 kW	AU$0.60	15 yrs	Began 2009
South Australia	Net	10 kW	AU$0.44 + retail price	20 yrs	Began 2008
Australian Capital Territory	Gross	200 kW	AU$0.30	20 yrs	Began 2009
Western Australia	Net				Finished 2011
Queensland	Net	10 kW	AU$0.44	20 yrs	Began 2008
New South Wales	Gross	10 kW	AU$0.66 (from July 2010)	7 years	Closed

Source: www.energymatters.com.au/government-rebates/feedintariff.php

Conclusion

Looking at the various policies and instruments that have been adopted by countries to promote solar power it is clear that there is not yet a single strategy that has been accepted as a mechanism to promote this technology. It is also far from clear that all countries accept that this is even a technology to be promoted. Despite all the different mechanisms though a few things are clear. Firstly, if the policy environment is right, then solar is a technology that can grow extremely quickly – so quickly in fact that a great deal of thought needs to be given to the level of support on offer in order to prevent support becoming unsustainable and leading to a boom and bust situation. The second thing is that solar is such a fast moving area that policies need to be flexible and up to date. Too much uncertainty is bad for any industry, however, solar technology is advancing so fast that governments frequently seem out of touch, reliant on false assumptions or out-of-date information. On the burning question of whether feed-in tariffs are better at stimulating growth than quota-based systems, it seems clear that for the majority of countries feed-in tariffs have the edge, particularly for the development of decentralised and domestic applications. In the future, as costs drop, quota systems may become more appropriate to drive prices down further and allow large-scale industrial expansion. In this respect both may be necessary. Feed-in tariffs introduce the technology to the market and establish expertise and economies of scale which reduce prices, allowing the fixed tariffs to be reduced and other mechanisms to take over. It also seems clear that it may not be long before support policies are no longer necessary as costs drop and the price of fossil fuels climbs.

Chapter 5

Existing and planned projects

Despite rapid growth and huge potential, concentrating solar power and photovoltaics are still energy sources in their infancy. In many countries PV is still largely thought of as a roof-top application, and I suspect that most people outside the energy and environment business have never really heard of CSP. The practical use of these technologies for large-scale applications has only just begun to enter the popular consciousness of much of the world, and to many they remain space-age or fanciful. This is understandable, since solar currently accounts for just 0.01 per cent of global electricity production.[1] Outside of key markets in Japan, China, California and Europe, many people may have never knowingly seen a solar system. The mere fact that this book can list all of the existing CSP plants in one table is indication enough of just how far things have to go. Were this book looking at the number of coal power stations or even wind farms, the list of installations could fill many pages.

Nonetheless solar power is an industrial sector that is growing fast and changing rapidly, and writing this text has been a constant struggle to keep on top of the latest developments. Each new plant is invariably hailed as the world's first or world's largest, while names of projects and companies change frequently, and developers swing between wanting to focus on one technology and then the next. What is true is that a number of multi-megawatt PV and CSP plants have already been built and several in the 100 megawatt class are already under construction. CSP in particular is going through a period of wild fluctuation. Just a few years ago it was seen as a leading candidate for large-scale energy in the desert, and in particular North America. Now many of the companies involved are reported to be struggling and rumours are circulating that many developers are swapping from CSP to PV. At the same time though large companies like GE and Areva have declared their interest in CSP, particularly in emerging markets.

Why has CSP gone through these ups and downs? There are several reasons. For many years it appeared as though it would offer a lower cost range than PV (roughly $0.10–0.20 per kWh), but in the last 2–3 years the price of PV has fallen making this much less certain, and undermining the business case for concentrating solar. At the same time the greater reliability of electricity supply available with CSP, thanks to heat storage and the ability to use natural gas as back-up, continues to make it attractive, as does the potential for cost reductions by localising manufacture.[2]

Aside from cost, CSP does have some other drawbacks compared with PV, particularly the need for water to cool the generators in some technologies and greater land requirements (see Chapter 2 on technology and also the environmental issues in Chapter 7). PV has other advantages too, such as its small number of moving parts (just the tracking system) and autonomy of operation, both of which could lead to lower downtime than a more complex CSP station. PV is not entirely self-standing though. Recent experiments in Abu Dhabi (see section on Masdar in Chapter 6) have shown how the build-up of desert sand and dust can reduce the power output of PV panels by up to 20 per cent in just four months, meaning the costs and energy inputs of frequent washing must be factored into the equation. Most importantly for PV though, there are real hopes that the cost of manufacturing could continue to fall significantly in the future, making it more competitive with mainstream sources of energy, and potentially allowing it to emerge as the cheapest source of energy.

Since relatively few large-scale CSP or PV plants are currently operational, this section will look not only at what has actually been built, but also at those projects which are in an advanced stage of planning, or which are under construction. When writing on topics such as this it is always hard to know just how many of the proposed projects will ever be commissioned, or exactly what stage developments are at, but the sections below provide a useful measure of the scale and scope of activity in large-scale solar.

Large-scale CSP installations today

Since they are essential to help with explanation, the first few concentrating solar thermal power stations are outlined in the chapter on technology, and indeed, in parabolic trough technology, the largest of the existing power stations is also one of the oldest. This is unlikely to be the case for very long, however, as now after years of stagnation, the technology is surging ahead, led by over 2000 MW of new plants in Spain, and up to 9 GW of projects in the pipeline in the USA (although as mentioned some of these may be switching to PV as costs in that sector fall). For the first time too, North African and Middle Eastern countries will soon have fully operational CSP plants, perhaps the first stage of much more ambitious projects.

CSP plants in the USA

The deserts of the South-Western United States can rightly lay claim to being the home of modern CSP development, and have been the key testing and development ground for much of the technology. Despite its early start, the USA has taken its time to re-engage with the concentrating solar renaissance, and currently trails a little behind Spain in terms of megawatts under construction. It seems likely that part of the reason for this slow start has been the lack of a guaranteed feed-in tariff for solar electricity. The USA operates a series of grants and tax incentives along with the renewable obligation system, whereby the amount of renewable energy required is set by the individual states, and the exact technology is left to the market. While in many ways this seems a sensible solution, which should encourage new and innovative forms of renewable energy, in practice it appears to promote slower growth than the alternatives, in the short term at least (see markets section in Chapter 3).

Thankfully, however, things in the USA are beginning to move, and there is a range of new market entrants and exciting projects in the pipeline. Aside from the original parabolic trough systems mentioned above, six systems – a 1.5 MW dish–Stirling system at Maricopa, New Mexico, a 5 MW power tower in California, a 5 MW Fresnel reflector and three separate parabolic troughs totalling 66 MW – have come into operation in 2005–2010, making the USA the only country with MW-scale working examples of each of the major CSP technologies. Following on from these initial projects, work is under way on a number of genuinely large installations, with a total of over 9500 MW of developments announced, from a range of technologies (see Table 5.1).

Interestingly, whereas the largest installations under development in Spain have a capacity of 200 MW, and are typically made up of smaller 50 MW units, those in the USA are far larger, up to 900 MW in size and averaging around 200–500 MW. The reason is probably because the cap for the Spanish feed-in tariff is set at 50 MW, making anything larger uneconomic. Plants with a capacity of more than 50 MW in Spain are made up of units of 50 MW arranged side by side. This therefore may be one example where the feed-in tariff may actually be preventing Spanish developers from making economies of scale in their investments.

The plans in the USA also call for a much broader range of technology. The deployments in Spain are overwhelmingly parabolic troughs, with a few small power towers and experimental dish–Stirling and LFR systems thrown in. The large installations planned for the USA are more evenly split between parabolic troughs, power tower, linear Fresnel and dish–Stirling systems. Again, over time, this should help to promote innovation and reduce costs to the most cost-effective technologies, although in the short term it could lead to delays due to reduced standardisation in the supply chain. Having said that, a number of small dish–Stirling and LFR plants are being planned for Spain, but these remain very much the exception to the rule.

Table 5.1 Large CSP installations in the USA

Name of plant	Location	Peak capacity	Technology	Status
SEGS I–IX	Mojave desert, California	354 MW	Parabolic trough	Operational
Nevada Solar One	Nevada	64 MW	Parabolic trough	Operational (2007)
Saguaro	Red Rock, Arizona	1 MW	Parabolic trough	Operational (2006)
Keahole Solar Power	Hawaii	1 MW	Parabolic trough	Operational (2009)
Sierra Sun Tower	Lancaster, California	5 MW	Power tower	Operational (2009)
Kimberlina Solar Thermal Energy Plant	Bakersfield, California	5 MW	Fresnel reflector	Operational (2009)
Maricopa Solar	Peoria, Arizona	1.5 MW	Dish–Stirling	Operational (2009)
Martin Next Generation Solar Energy Centre	Florida	75 MW	Parabolic trough – ISCC	Operational

Parabolic trough systems in the USA

The USA is home the world's first and largest parabolic trough systems in the form of the nine Solar Energy Generation Systems (SEGS I–IX) at Kramer Junction in California (Figure 5.1). These have a combined capacity of 354 MW and continue to operate well. While the company which built them – Luz Engineering – has gone out of business, they are now run by NextEra Energy Resources and Cogentrix Energy and, combined, produce around 654 GWh per year, at an estimated cost of around $0.12–0.14/ kWh,[3] although some commentators suggest this has fallen to around $0.09 / kWh.[4]

After many years as a lone symbol, these plants are now being joined by the new wave of parabolic troughs. Work on these projects has been slow to get going though, and in the last five years the USA has installed just over 66 MW of new parabolic trough capacity, consisting of a 1 MW parabolic trough plant at Saguaro in New Mexico, a 1 MW facility in Hawaii and the 64 MW Nevada Solar One in Nevada (see Table 5.1).

Completed in 2006, the Saguaro plant was the first megawatt-scale parabolic trough system to be constructed anywhere in the world since the last of the SEGS systems in 1990. Covering an area of 16 acres (6.4 hectares), the plant operates without back-up or storage and produces around 2000 MWh per year for owner and operator Arizona Public

Figure 5.1 SEGS IV parabolic trough power plant in California, USA (source: DOE/NREL/ Sandia National Laboratories).

Services. As with previous troughs, the mirrors for the Saguaro plant were produced by Flabeg of Germany, and the plant utilises an organic Rankine turbine, in which liquid organic fluids such as pentane are used as the working fluid.

Moving up in size, the 64 MW Nevada Solar One plant was developed by Solargenix (currently owned by Acciona Energy of Spain) and became operational in 2007. Costing a total of $260 million to construct, the plant covers an area of roughly 400 acres (161 hectares) and generates over 130,000 MWh per year, enough to power around 26,000 homes in the region. Like the Saguaro plant it operates without any fossil fuel backup, although it does employ thermal storage with sufficient capacity for 30 minutes of operation.

Finally, 2009 saw the completion of a third parabolic trough facility, this time on the Big Island of Hawaii. Constructed for Keahole Solar, a subsidiary of Sopogy, the 1 MW facility is the first of up to 30 MW the company hopes to develop on Hawaii, reducing the island-state's energy imports. Hawaii is a potentially attractive location for renewable energy developers, since it is the most expensive electricity market in the United States (in 2010 residential prices were around $0.29 / kWh compared with $0.15 / kWh in California).[5] This means that CSP has potential to be directly competitive with traditional forms of power generation.

These three initial projects are the tip of the parabolic iceberg, and there are presently applications filed for a further 4350 MW of parabolic troughs right across the country. Of these, the 75 MW Integrated Solar and Combined Cycle (ISCC) plant being developed for Florida Power and Light (FPL) in Martin County, Florida, looks likely to be among the first to be completed, and is the only one to have so far broken ground. Construction on the $476 million plant began in 2009 and was finished in 2010.[6] Once completed the plant will be the first in the world to attach a solar field to an existing combined cycle gas generation facility.

Of the larger parabolic trough systems in the pipeline, one of the most advanced is the 280 MW Solana plant, near Gila Bend in Arizona. The project received a huge boost in July 2010 when the company constructing it – Abengoa Solar, a US division of Spanish giant Abengoa – was awarded $1.4 billion in US loan guarantees by President Obama. According to the developer the plant will cover an area of 1900 acres and provide electricity for up to 70,000 homes. The power will be sold to the Arizona Public Service (APS).[7]

Looking further into the future, it is hard to predict which of the parabolic trough systems will be next to be constructed, or even how many are likely to see the light of day. In six years of writing about renewable energy projects, I have learned that delays are the norm, and that developers will claim that everything is on track until the very last minute, when suddenly the plug can be pulled. What *is* clear is that there is a tremendous amount of interest in this type of technology, with key desert areas in California, New Mexico and Nevada in particular emerging as hotspots for CSP development. What happens then in the Sonoran and Mojave deserts will have profound implications for the rest of the world, and developments are being followed keenly by engineers and policy makers alike.

It is important to remember what a crucial and formative period this is for solar power, and for CSP in particular. While one technology may look to be an interesting and exciting project, with the potential for real cost reductions in the production of solar electricity, this field is moving so fast that it is entirely possible that some other innovation or design will come along in the next two to three years. In the coming decade the whole industry is likely to go through a shakedown process, with new ideas coming to the fore and a range of radical ideas being tested before a standard design begins to take shape. While power towers and parabolic troughs have the early advantage, other concentrating systems are also under development (Table 5.2).

At present, the United States has only one functional power tower system, the 5 MW Sierra SunTower in Lancaster, California. Completed in 2009, this plant is owned and operated by eSolar, and to date is the only working multi-megawatt commercial power tower outside of Spain. Unlike the Spanish plants, however, the Sierra SunTower uses no fossil fuel back-up or heat storage system. Neither does it receive a feed-in tariff, instead relying on the 30 per cent Investment Tax Credit to reduce capital costs.

Covering an area of 20 acres (8 hectares), the SunTower is made up of two separate modules, each consisting of its own tower and thermal receiver, supplied by 12,000 mirrors. The electricity produced is being sold to Southern California Edison.[8]

eSolar's principal innovation is in the design, manufacture and control of its heliostats. By modifying the shape of the mirrors and reducing their size,

Table 5.2 Large CSP installations under development in the USA

Name of plant	Location	Peak capacity	Technology	Status
Solar Millennium Blythe	Kern County, California	1000 MW	Parabolic trough	Approved
Calico Solar (formerly SES Solar One)	San Bernardino, California	663–850 MW	Dish–Stirling	Approved
Imperial Valley (formerly SES Solar Two)	Imperial Valley, California	709 MW	Dish–Stirling	Approved
Mojave Solar Park	San Bernardino, California	553 MW	Parabolic trough	Unknown
Fort Irwin	San Bernardino, California	500 MW	Unknown	Unknown
Ivanpah Solar Power Tower	Kern County, California	370 MW (nominal), 393 MW gross	Power tower	Approved
Hualapai Valley Solar Project	Mohave, Arizona	340 MW	Parabolic trough	Unknown
Agua Caliente Solar Project	Yuma, Arizona	290 MW	Parabolic trough	
Solana Generating Station	Gila Bend, Arizona	280 MW	Parabolic trough	Recently awarded $1.45 billion in loan guarantees
Beacon Solar Energy Project	Kern County, California	250 MW	Parabolic trough	Approved
Solar Millennium Palen	Kern County, California	484 MW	Parabolic trough	Approved
Abengoa Lake Solar	San Bernardino, California	250 MW	Parabolic trough	
Amargosa Solar Power Project	Amargosa Desert, Nevada	250 MW	Parabolic trough	
Project Genesis	Riverside, California	250 MW	Parabolic trough	Approved
Solar Millennium Ridgecrest	Kern County, California	250 MW	Parabolic trough	Under review
Kingman Solar Project	Mohave County, Arizona	200 MW	Parabolic trough	
Rice Solar Energy Project	Riverside County, California	150 MW	Power tower	Approved
City of Palmdale Hybrid Gas-Solar	Palmdale, California	62 MW solar 555 MW gas	Parabolic trough / gas hybrid	Under review
Victorville 2 Hybrid Power Project	Victorville, California	50 MW 513 MW gas	Parabolic trough/gas hybrid	Approved

Source: The California Energy Commission http://www.energy.ca.gov/siting/solar/index.html

the modules can be more easily produced and assembled, and packed into shipping containers. Meanwhile with smaller mirrors the mounting systems can be simplified as they benefit from less wind resistance. To control all these new, smaller mirrors (there are many more small mirrors than in traditional power tower designs) the company has needed to develop advanced software that enables it to more efficiently focus sunlight onto its thermal receivers. As a result of all of this the company claims it will be able to install fields more quickly and cheaply compared with other CSP developers, and requires less skilled labour. While it is still early days, eSolar claims to have developed partnerships with several overseas companies and is looking to supply towers and mirror fields for projects in Turkey and China. The company has also lined up an impressive array of investors including Google, NRG and Idealab.[9]

Supplying 4000 households is a modest start, but the next wave of power tower developments in the USA is likely to be far larger. The 370 MW Ivanpah project, located near the town of Primm on the California–Nevada border, is just outside the Mojave Desert wilderness area. Under development by Brightsource Energy using its Luz Power Tower technology (see boxed text), the project shot to fame after being awarded a $1.37 billion grant by the US Department of Energy. Construction on the project is due to start in 2010 with the first 100 MW phase operational by 2013.[10,11] According to some sources, once complete the plant should have a capacity factor of around 30 per cent and be able to produce electricity at a price that is almost competitive with conventional generation – coming in at less than $0.125 per kWh,[12] and making it not far off being competitive with natural gas or coal.[13]

Stirling engine systems

To date the USA is alone in having advanced plans for the deployment of large Stirling engine systems. Unlike parabolic troughs, power towers or Fresnel reflectors, Stirling systems do not use standard thermal generators and condensers, and so have minimal water requirements. At the same time, their direct electricity production precludes any form of heat storage system. It remains to be seen which of these will turn out to be the most crucial element in the coming years. For now though there is only one large example of a Stirling engine system, the 1.5 MW Maricopa facility at Peoria, Arizona. This is very much a demonstration project, and was the first time that Stirling Energy Systems (SES) and Tessera Solar (the only commercial manufacturers and distributors of this technology) were able to fully show off their modular, 25 kW Suncatcher units. Previous to this the largest array had been six experimental systems constructed at the Sandia National Laboratories in New Mexico, also by SES.

The next stage for Stirling engine technology should involve a dramatic increase in the scale of deployment. In 2010, construction is due to begin on the first stage of the Calico Solar project (formerly SES Solar One). Once completed, this would have a capacity of 663 MW and cover an area of 8000 acres (3237 hectares) of desert, some 37 miles (57 km) east of Barstow, California. As is normal with this type of development, Calico Solar will be constructed in stages, the first around 300 MW and made up of 1200 Suncatcher systems.[14]

The second of Tessera Solar and SES's large developments is Imperial Valley (formerly the imaginatively titled SES Solar Two). Sprawling over 6500 acres (2630 hectares), this 750 MW development will also be built in two stages: 300 MW and 450 MW. In order for the second half of this project to be completed, the developers are counting on the construction of the Sunrise Powerlink, a 500 kV grid extension which will run about ten miles from Solar Two.[15] Without this extension, there will only be sufficient capacity in the existing grid for the first 300 MW development. The reliance on expensive grid infrastructure, even in as developed a place as California, gives a taste of some of the hurdles which will need to be overcome in even more remote, and in some cases, far poorer locations.

Linear Fresnel reflectors

At present there is just one multi-megawatt linear Fresnel reflector CSP plant operational in the United States. Located at Bakersfield California, the 5 MWe Kimberlina plant is the first of its kind in the USA and was constructed by Ausra using collectors manufactured at a plant in Las Vegas. The plant is actually the second to have been built by Ausra, with the first facility in Australia attached to a coal power station, in which the steam generated by the solar plant is used to offset a small portion of that generated by coal. Since it was constructed, Ausra has been taken over by French giant AREVA which is pursuing the technology, with a reported pipeline of up to 1000 MW of CSP systems.[16]

CSP plants in Spain

Though not strictly a desert region, Spain does receive a steady and predictable supply of sunshine, with an average solar irradiation of 1200 kWh/m^2 to over 2000 kWh/m^2 depending on region.[17] Combined with a favourable policy environment of fixed feed-in tariffs, it has emerged as an undisputed global leader in the deployment of CSP technology. It was the first country to build commercial systems and now has an impressive pipeline of projects. In fact, in many ways it is to CSP what Germany has been to PV. Although neither country has an absolutely ideal resource, a mixture of favourable policy and economic conditions has allowed them to become global leaders in renewable technologies. It remains to be seen if each will be able to keep its leading position in the coming decade, particularly in light of the

Box 5.1 **Ivanpah Solar Tower Project**

The 393 MW Ivanpah Solar Electric Generation System in California's Mojave desert is one of the largest and most high-profile Concentrating Solar Power stations currently under development. Perhaps the main reason for this is that Brightsource Energy is in many ways the heir of Luz International, the developer of the original SEGS systems, and as such it has a strong track record in CSP technology. This has helped Brightsource to attract the backing of high profile companies like Google and BP, and to secure a loan of $1.37 billion from the US Department of Energy.

Founded in 2006 by an amalgamation of Luz II (headed by Arnold Goldman, the founder of the original Luz International) and a group of investors, Brightsource Energy has been testing its power tower technology in Israel's Negev desert since 2004, and is now set to introduce it to California. The very fact that the company has chosen to focus on power tower technology, rather than sticking with parabolic troughs (like the SEGS) is in itself interesting. According to Brightsource, the experience of working with parabolic trough systems has also highlighted that technology's limitations, in particular its inability to track the sun in two axes and the inefficiency of 'moving' the heat such long distances through the field of troughs. In response to these perceived limitations, Luz II was founded in 2004 to develop and advance new power tower technology.

The new Ivanpah plant, at Kern County in California, will be constructed in three phases. The first two phases will have a generation capacity of 100 MW each and will each consist of three towers and receivers and three fields of heliostats. The third phase will have an output of 200 MW and will be made up of four sets of towers and heliostats. Each will be built next to the other to allow the installation as a whole to benefit from shared access to the grid, as well as roads and infrastructure. According to the California Energy Commission the whole site will occupy around 5.3 square miles (13.7 km^2).[18]

The Ivanpah project is a good example of how CSP plants are already being forced to seriously consider their water demands. Once completed the power will use a dry-air cooling system, purportedly reducing its water demands for cooling by around 90 per cent. The company also claims that by mounting the heliostats on poles, they reduce the amount of land that needs to be 'levelled' with benefits for plants and wildlife.

As is the case with many CSP developments, natural gas will play an important role in the installation too, providing back-up generation and helping to increase the running time to about 10 hours a day.

Table 5.3 CSP projects operational in Spain, end 2010

Name	Technology	Capacity	Location	Completed
Andasol 1 + 2	Parabolic trough	100 MW	Guadix, Granada	2008, 2009
Alvarado 1	Parabolic trough	50 MW	Badajoz	2009
Extresol 1 & 2	Parabolic trough	100 MW	Badajoz	2009, 2010
Solnova 1,3,4	Parabolic trough	150 MW	Seville	2010[a]
Ibersol Puertollano	Parabolic trough	50 MW	Ciudad Real	2009[b]
La Florida	Parabolic trough	50 MW	Badajoz	2010[c]
Madajas de Tietar	Parabolic trough	50 MW	Badajoz	2010
La Dehesa	Parabolic trough	50 MW	Badajoz	2010
Palma de Rio 1	Parabolic trough	50 MW	Cordoba	2010
Manchasol	Parabolic trough	50 MW	Caceres	2010
La Risca	Parabolic trough	50 MW	Badajoz	
PS10	Power tower	11 MW	Seville	2007
PS20	Power tower	20 MW	Seville	2009
Peurto Errado i	Linear Fresnel reflector	1.4 MW	Murcia	2009

a *Abengoa Press Release* 22 September 2010 – Abengoa Solar Reaches Total of 193 Megawatts Operating http://www.abengoasolar.com/corp/web/en/about_us/general/news/archive/2010/solar_20100 802.html
b *Iberdrola Press Release* http://www.iberdrolarenovables.es/wcren/corporativa/iberdrola?IDPAG =ENMODULOPRENSA&URLPAG=/gc/en/comunicacion/notasprensa/090508_NP_primera_ termosolar.html
c *Energy Matters* 15 July 2010 'Spain – A solar thermal powerhouse.' http://www.energymatters.com. au/index.php?main_page=news_article&article_id=960

Source: Protermo Solar – Spanish Solar Thermal Industry Association http://www.protermosolar. com/boletines/32/mapa.html

recent global financial crisis and government belt-tightening that is going on. Regardless of their domestic markets though, it seems likely that both are poised to play a major role in exporting technology and skills as other markets begin to catch up, and Spanish companies like Abengoa are already offering their expertise in concentrating solar power as far afield as the USA, China and the Persian Gulf.

In terms of what has already been constructed, Spain has around of 750 MW of parabolic troughs including the 100 MW Andasol Solar Power Station (made on Andasol-1 and Andasol-2) and 13 other 50 MW installations that were completed in 2009 and 2010 (Table 5.3). One of these (Solnova) forms part of the Solucar solar complex, an interesting 300 MW facility that is being developed by Abengo. This appears to be a concerted effort to try and test out each of the various CSP technologies. While most of its final 300 MWe will come from parabolic trough generators, it also includes two functioning power towers, PS10 and PS20, which together have a combined capacity of 31 MW, and will host a small linear Fresnel reflector and a 60 kW dish–Stirling system.

Added to this growing clutch of operational systems is an impressive pipeline of plants already under construction or in the pre-construction phase (see Table 5.4). Indeed, of the CSP actually being physically built at the time of writing, the vast majority is in Spain.

In contrast to the United States, however, Spanish developers have largely opted for a single technology – the parabolic trough. Only about 5 per cent of the CSP plants being developed use any other technology, principally power tower followed by a few Fresnel reflector systems and an 80 kW dish–Stirling project planned for the Solucar complex. Another 74 MW of dish–Stirling systems are under consideration near Puertollano. It could be that the rush to construct and commission projects before any further changes are made to the Spanish feed-in law is encouraging developers to rely on the most mature CSP technology available, and not to take too many risks with technology.

Parabolic troughs

Altogether, just over a dozen parabolic trough systems have been completed and commissioned in Spain. Located in the foothills of the Sierra Nevada mountains near Guadix in Granada, the first two of these, Andasol I and Andasol II, were developed by Solar Millennium and were commissioned in 2009. When it was completed, Andasol I was the first large parabolic trough system in Spain and the first outside the USA.

Covering 200 hectares each, Andasol I and II are each expected to generate around 160 GWh per year, at a cost of around €0.271 / kWh.[19] As explained in the technology section, both these plants employ molten salt storage systems, with sufficient storage capacity for 7.5 hours of operation.[20]

The proximity of the mountains is a potentially important feature of the Andasol projects since they are located in an area of above average water availability. This means that their annual demands for 870,000 m^3 of water for cooling (4.8 m^3 per MWh) will be more easily met than they might be in other areas of Spain.

Andasol I and II are just the first stage in the Guadix development, and two more 50 MW parabolic trough power stations – Andasol III and IV – are also under construction, with completion estimated for 2011–2012.

Aside from the first two Andasol developments, around a dozen other parabolic trough projects have been completed in Spain. These are listed in Table 5.4. Like the Andasol projects, these are 50 MW developments,

Table 5.4 List of CSP projects reportedly under construction or in pre-construction in Spain, March 2010

Name	Technology	Capacity	Location	Notes
Extresol 3	Parabolic trough	50 MW	Badajoz	7.5 hour heat storage
Andasol 3 + 4	Parabolic trough	100 MW	Guadiz, Granada	7.5 hour heat storage
Palma de Rio 2	Parabolic trough	50 MW	Cordoba	
Helioenergy 1 + 2	Parabolic trough	100 MW	Ecija	7.5 hour heat storage
Solaben 1 + 2	Parabolic trough	100 MW	Logrosan	
Valle Solar Power Station	Parabolic trough	100 MW	Cadiz	7.5 hour heat storage
Aste 1A + 1B	Parabolic trough	100 MW	Alcazar de San Juan	
Termosol 1 + 2	Parabolic trough	100 MW	Badajoz	
Helios 1 + 2	Parabolic trough	100 MW	Ciudad Real	
Lebrija 1	Parabolic trough	50 MW	Lebrija	
Axtesol 2	Parabolic trough	50 MW	Badajoz	
Arenales PS	Parabolic trough	50 MW	Seville	
Serrezuella Solar 2	Parabolic trough	50 MW	Badajoz	
El Reboso 2	Parabolic trough	50 MW	Seville	
Moron	Parabolic trough	50 MW	Seville	
Olivenza 1	Parabolic trough	50 MW	Badajoz	
Medelin	Parabolic trough	50 MW	Badajoz	
Valdetorres	Parabolic trough	50 MW	Badajoz	
Badajoz 2	Parabolic trough	50 MW	Badajoz	
Sanata Amalia	Parabolic trough	50 MW	Badajoz	
Torrefresneda	Parabolic trough	50 MW	Badajoz	
La Puebla 2	Parabolic trough	50 MW	Seville	
Termosolar Borges	Parabolic trough	25 MW	Lerida	
Gemasolar/ Solar Tres	Power tower	17 MW	Seville	15 hour heat storage
Renovalia	Dish–Stirling	1 MW	Albacete	

Source: http://en.wikipedia.org/wiki/List_of_solar_thermal_power_stations#cite_note-Acciona_2-20; http://www.protermosolar.com/boletines/32/mapa.html

and are the first stages of larger generating facilities. The exact number of completed plants is hard to be sure of as it changes very fast, with nearly ten new projects coming online while the research for this book was being conducted. The key point is that this is a rapidly developing field, with huge investment continuing to flow into the sector.

Looking at the next wave of projects under construction, it seems that 50 MW is very much the unit of CSP construction in Spain, with every single parabolic trough under development consisting of one or more 50 MW generators. The reasons for this have already been mentioned and likely have to do with the structure of the Spanish feed-in tariff that sets a cap of 50 MW on the size of any one development. While it has come in for criticism, the tariff has at least been successful in fulfilling its primary objective – that of increasing the amount of electricity produced from solar technology. With around 1600 MW of parabolic troughs under construction, CSP looks set to make a significant contribution to Spanish electricity needs in the coming years. In fact, if the nearly 3000 MW of parabolic troughs is each assumed to have the same output as Andasol I (160 GWh year/50 MW capacity), Spain should be able to produce roughly 9.6 TWh of electricity from CSP, much of it at times of peak demand. At 2009 consumption this represents nearly 3.5 per cent of total electricity consumption.[21] Combined with solar PV and wind, renewable energy should be consistently providing nearly 30 per cent of the nation's electricity in the next few years.

Power towers

Compared to the USA, where they are proving a popular technology, power towers are currently a fringe concern in Spain. While it is true that two have recently been completed (PS20 and PS10, see Figure 5.2), there is currently only one more under construction, the 17 MW Solar Tres. Once these three are completed, they will have a combined capacity of only 48 MW, compared with thousands of MW of parabolic troughs in the works.

CSP under development in other markets

While the USA contains vast areas of true desert, and Spain has an excellent solar resource, neither quite fulfils the image that this book lays out, one of vast grids of solar generators in the sand, churning out endless clean electricity. There are two main reasons for this. The first is that developments in these two countries are market driven, decentralised, and messy, which does not lend itself to the imagination in quite the way that planned super-grids do. The second is that both these countries are highly developed. While this makes the technology no less revolutionary or beneficial, it does not perhaps have the same psychological impact that a few square kilometres of India or Egypt might.

Figure 5.2 The Solucar complex in Spain includes the PS10 and PS20 power towers (source: Abengoa).

To date, countries outside of North America and Europe have been slow to adopt large-scale solar, held back by the high costs, uncertainty over the new systems and technical barriers. Despite this a number of other countries *do have* advanced plans for concentrating solar power, particularly in North Africa. Currently Algeria, Morocco and Egypt are leading the way, with three projects under construction that contain a concentrating solar power element. Without exception the solar element consists of a parabolic trough system, reinforcing this technology's status as the most tried and tested form of CSP.

At Hassi-R'mel in Algeria, work is under way on what would have originally been the world's first integrated solar and combined cycle system (ISCCS). Backed by a 2004 law that introduced a feed-in tariff for CSP installations (currently set at 100–200 per cent above standard electricity prices depending on the percentage of solar energy a power station utilises), this 150 MW development consists of a combined cycle gas generator, and a 20 MW parabolic trough system. Financed by Abengoa and NEAL (New Energy Algeria), the project is offering a template for other developments in the region. Once completed, the aim is for Abengoa and NEAL to operate the plant for 25 years, with the electricity sold to the local market.[22]

Next door in Morocco, Abengoa is building another ISCC system for the Office Nationale d'Electricité (ONE). The project will consist of a 450 MW ISCC system which is to include a 20 MW CSP component. While the solar part of this installation seems disappointingly small, projects like this could help provide an important transition for an emerging industry. Indeed this was very much the original idea when the project was designed. The World

Bank, which was one of the key funders of the project, was concerned that solar was too new at the time, and that this represented a useful introduction for the region. Construction contracts for the Moroccan development were awarded to Abener in 2007 and work is proceeding.[23]

More widely, the Moroccan government seems to be recognising the importance of renewable energy and has announced it will invest $3.2 billion in a five-year renewable energy project which will run until 2014. As part of this they aim to achieve 10 per cent of energy supply from renewables by 2012, up from 4 per cent in 2008. Part of this investment will also fund a new campus for renewable energy and high technology research, but already calls have been put out for companies to construct a 500 MW CSP plant.[24,25]

In Egypt, meanwhile, construction finally seems to have begun on the long-awaited Kuraymat CSP development, located 90 km south west of Cairo. While this projected has been in the pipeline for many years, only recently has it seen on-the-ground progress. Financed by a loan from the UN's Global Environment Fund (GEF) and the New and Renewable Energy Authority of Egypt (NREA), Kuraymat Solar consists of a 146 MW ISCC system with a 25 MW parabolic trough CSP component. Under development by Spanish renewables giant Iberdrola and Orascum/Flagsol, the project became operational in summer 2011 and will become Egypt's first large-scale solar development.

Farther to the east, 2009 saw the completion of Iran's first concentrating solar thermal power station in Yazd, at the time the world first solar and gas integrated combined cycle system. With a total capacity of 414 MW and a solar component of 67 MW covering an area of 366,240 m^2, the plant is one of the largest solar thermal systems currently operational. The total solar contribution of the solar field is estimated to be 5.3 per cent over the course of the year.

Israel too is seeking to exploit the potential of its deserts. Despite its companies and researchers playing a leading role in the development of concentrating solar technology (through Solel and Luz Engineering), Israel currently has no large-scale solar power plants of any kind. This should all be about to change, however, with three projects under construction in the Negev Desert, at Ashalim, and the Samar and Yavne Kibbutzim.

The development of the Ashalim project began as far back as 2003 when the Israeli government introduced a feed-in tariff for solar electricity, set at $0.163 / kWh for CSP installations of over 20 MW, and $0.204 for those which are smaller. As part of this review, plans were drawn up for an initial CSP plant of 100 MW, with space to expand it to 500 MW. In 2006, a 400 hectare (1000 acre) site at Ashalim was selected. For the next four years there was a long pause, but finally in 2010 a tendering process was announced to develop an integrated renewable energy installation at the site, consisting of 2 × 120 MW CSP installations and 15–30 MW of PV. The total cost of the facility is estimated to be around $750 million, with the European

Investment Bank providing up to €100 million in loans. At present it is not clear what technology will be successful but with major Israeli companies involved in the process it could be either power tower or parabolic trough technology. The whole project is expected to be complete in 2014, and once fully operational, the combined CSP and PV output could provide 2.5 per cent of Israel's total electricity supply.

Aside from the large Ashalim plant there are plans under way for a number of smaller decentralised CSP plants throughout Israel. The first of these was constructed by local company AORA at the Samar Kibbutz near Eilat in the south of the country, following the construction of a prototype near Nanjing in China. Consisting of a small power tower with natural gas back-up spread across 0.5 acres, the plant has a capacity of 100 kWe and 170 kWth. Plans are afoot to offer the technology to other kibbutz and small communities, both in Israel and in Spain. Already at the time of writing, a second kibbutz at Yavne has announced it will build a small CSP installation at a ceremony attended by the Israeli President Shimon Peres.

Large-scale photovoltaics today

While CSP has been claiming most of the headlines, photovoltaics too has seen a dramatic increase in the number and scale of multi-megawatt developments over the last few years. Centred largely on Germany, Spain and the USA, with other developments in Italy, South Korea, China, Czech Republic, Canada and Portugal, there are now hundreds of multi-megawatt plants spread across the world. Of these, Spain, Italy and Germany have accounted for by far the largest number of large-scale PV installations, and this is reflected in the list of largest PV power stations in Table 5.5. In fact by the middle of 2010, Europe as a whole accounted for around 85 per cent of all the large-scale solar currently employed worldwide (greater than 200 kW). Because of this, as elsewhere in this book, it will be impossible to describe the current state of large-scale PV installations, and so illustrate their potential for deployment in the desert, without looking at some of the key European markets.

In terms of technology, large-scale PV remains dominated by crystalline silicon, which accounts for about 85 per cent of the market (compared with an 80 per cent share of the PV market as a whole). Of the thin-film technologies, cadmium telluride is the largest, with an 8 per cent share of the market. As large-scale PV moves into the deserts this technology mix is likely to change. The huge amounts of land available in desert regions, the greater performance of thin-film technologies under high temperature and rapidly falling costs could make the economics of large-scale thin-film PV more attractive. For now though, crystalline silicon remains dominant (see Chapter 2).

When writing about PV it can be harder to get an idea of what is happening than with CSP, particularly when it comes to future projects.

Table 5.5 Largest PV systems in the world, February 2011

Power	Location	Constructed
97 MWp	Sarnia, Canada	2010
84.2 MWp	Montalto Di Castro, Italy	2010
80.7 MWp	Finsterwalde, Germany	2010
70 MW	Rovigo, Italy	2010
60 MWp	Olmedilla, Spain	2008
54 MWp	Strasskirchen, Germany	2009
53 MWp	Turnow-Preilack, Germany	2009
50 MWp	Puertollan, Spain	2008
48 MWp	Boulder City, USA	2010
46 MWp	Moura, Portugal	2008
43 MWp	Cellino San Marco, Italy	2010
40 MWp	Waldpolenz/Brandis, Germany	2007/2008
36 MWp	Reckahn, Germany	2010

Source: www.pvresources.com

This is partly because the decentralised nature of PV installations simply means that there are too many possibilities to keep track of. The shorter lead-in time for constructing a PV plant also means that they can seem to spring from nowhere. This is particularly the case in Europe, where there are a great many small multi-megawatt plants of around 5–10 MW being built each year. The boom and bust nature of the PV industry in Spain and the uncertainties surrounding the future level of the feed-in tariffs in other countries also mitigate against a clear vision of the future. As far as growth in scale is concerned, the general trend is definitely up, at least as far as the largest plants are concerned, and this is where the regions outside Europe will probably take the lead. In the USA for example, plants of several hundred megawatts each are currently on the drawing board, far larger than anything currently proposed for Germany, Spain or Italy. In China too there are grand government-driven plans for large-scale plants. While some of these are still long-term visions, and will be discussed elsewhere, for others they are tantalisingly close to realisation.

Spain

As of 2010, Spain has by far the largest amount of large-scale PV currently operational. By the end of 2008, it had at least 2290 MW of PV installations greater than 1 MW, most of which was developed in just two years between

2007 and 2008. A look at a list of large-scale PV installations in February 2011 illustrated this,[26] with 86 of the 200 largest PV installations in the world found in Spain. Of these at least 60 are greater than 10 MWp, totalling 984 MW.[27] Indeed the surge in large ground-mounted arrays was central to the story of the collapse of the Spanish PV market in 2009 (see Chapter 4). Since then the market has declined, from a high of 2290 MW of new large-scale solar plants in 2008 (over 200 kWp), to just 214 MW in 2009.[28]

With the reduction in the Spanish feed-in tariff and the shift in emphasis towards more decentralised and domestic systems it seems unlikely that multi-megawatt PV in Spain will continue to dominate as it has, and the boom in ground mounted installations will have moved on to the next country. Nonetheless the speed with which it was deployed is an indication of just how quickly this technology can be deployed when the economics are in its favour. Should the cost of PV ever be brought low enough to make it competitive without feed-in tariffs or other forms of support, we can expect rapid and explosive development across the world. Indeed, in the Spanish rooftop and domestic market, growth is already healthy, with more than 600 MW expected to be installed in 2010.

There are too many large PV installations in Spain to discuss individually, and a more complete list of projects is available online from www.pvresources.com.

Other European countries

Along with Spain, Germany, Portugal, Czech Republic and Italy have emerged as important potential markets for large-scale solar in Europe. Portugal in particular is currently home to one of the world's largest PV installations – the 43 MW Moura power plant. Covering roughly 250 hectares in the south of the country, the plant was constructed by Acciona energy and the first 43 MW phase was completed in 2008. An additional 20 MW plant is under construction at the site and was due to be commissioned by the end of 2010. In order to supply panels for the project, Acciona constructed a factory on site, providing polycrystalline silicon modules for the plant.[29]

Like its other Mediterranean neighbours, Italy also has an excellent solar resource (up to 2000 W/m^2/d), particularly in the south, and has constructed several multi-megawatt PV installations. The largest of these, and currently the second largest PV plant in the world, is the 84 MW Montaldo di Castro PV power plant in region of Lazio. Constructed by SunPower using that company's monocrystalline silicon panels and trackers, the project is one of a number of multi-megawatt projects that have been completed in Italy in 2009 and 2010.

The Czech Republic too has emerged as a large PV market, thanks to its generous feed-in tariff, with a total of 444 MW installed in 2009. At present the biggest plants are a 35 MW installation at Veprek, followed by a 14.2 MW installation at Ralsko and a 13.6 MW facility constructed by German company SAG Sonnarstrom at Stribo.[30] Whether the Czech Republic will remain an important market for solar will depend largely on the maintenance of its feed-in tariff, and there are already concerns that this will need to be adjusted to ensure that it is not providing unreasonable payback.

Despite its high latitude and relatively poor insolation, Germany is also a world leader in large-scale photovoltaics, just as it is in decentralised solar, thanks to its favourable policy environment. As with Spain there are too many large PV plants to list, but there are currently more than 1118 MW of PV installations in Germany made up of plants with a capacity of more than 1 MWp, and there are at least 19 plants with a capacity of over 10 MW.[31]

As the German government seeks to reduce the feed-in tariff payable to new PV installations, it remains to be seen if these sorts of large-scale installations will continue to thrive.

France and Belgium also have a number of large-scale PV facilities, including a 13 MW plant in Belgium and several generators over 5 MW in France, including a 10.5 MW PV installation on the French overseas territory of Reunion Island.[32]

Large-scale PV in the United States

In contrast to Spain and some other European countries where the boom in large ground-mounted PV systems appears to be drawing to a close, in the United States it is just beginning. More than 1700 MW of large PV is in the planning or construction phases, some in very large arrays of up to 600 MW (see Table 5.6). Interestingly, some of these large systems in the USA are using thin-film solar (for example the CdTe panels produced by First Solar, or Signet Solar's thin-film silicon panels). This may be a function of the greater area of land available in the USA compared with Europe, meaning that the lower module efficiency of the thin-film systems is less of a factor. It is also likely to be a reflection of the rapidly falling costs of these technologies, many of which are based in the USA.

Until these new projects are completed though, the largest PV installations in the United States are the 48 MW Copper Mountain plant in Colorado constructed by Sempra Generation and the 30 MW Cimarron plant in New Mexico (Figure 5.3), both completed in 2010 using CdTe panels from First Solar. These follow on from the 25 MW system at the DeSoto Next Generation Energy Centre and the 21 MW Blythe Solar Electric power plant. The first PV generator in the USA with a capacity greater than 10 MW, was the 14 MW installation at Nellis Air Force Base which was completed in 2007. Occupying 140 acres it should produce around 25 GWh, one quarter of the energy used on the base. In fact the enthusiasm which the US military has shown for renewable energy, and solar in particular, as a means

Table 5.6 Large-scale PV plants completed or planned in the USA (March 2011)

Name of plant	Location	Peak capacity	Annual output	Technology	Status
Copper Mountain	Boulder, Colorado	48 MW		CdTe	Operational (2010)
Cimarron	New Mexico	30 MW		CdTe	Operational (2010)
DeSoto Next Generation Solar Energy Center	Florida	25 MWe		Crystalline silicon	Operational (2009)
Blythe Solar Electric Power Plant	Blythe, California	21 MWe			Operational (2009)
Blue Wing	San Antonio, Texas	16 MW			Operational (2009)
Jacksonville SEG	Jacksonville, Florida	15 MW	25 GWh	Crystalline silicon	Operational (2010)
Nellis solar power plan	Las Vegas, Nevada	14 MWe			Operational (2007)
Alamosa PV plant	South Colorado	8.2 MW	17 GWh	Crystalline silicon	Operational (2007)
Rancho Cielo Solar Farm	Belen, New Mexico	600 MW	Unknown	Amorphous, thin-film silicon	Planning
Topaz Solar Farm	Carrizo, California	550 MW	1.1 TWh (estimated)	CdTe thin film	Planning
High Plains Ranch	California	250 MW	550 GWh (estimated)	Crystalline silicon, tracking	Planning
AV Solar Ranch One	California	210 MW	Unknown	Unknown, tracking	Planning
KCRD Solar Farm		80 MW	Unknown	Unknown	Planning
Davidson County Solar Farm	North Carolina	20 MW	36 GWh	Crystalline silicon	Planning

Source: PV Resources, 1 March 2011. www.pvresources.com

Figure 5.3 The 30 MW cadmium telluride Cimarron PV plant in New Mexico, USA (source: First Solar).

of hedging against fossil fuel costs and shortages, has probably gone some way towards promoting these technologies in the United States.

The 25 MW DeSoto plant in Florida was inaugurated by US President Obama in 2009. Using crystalline silicon panels and incorporating two-axis suntracking, the plant should provide enough electricity for 3000 of Florida Power and Light's (FPL) customers.

China

Currently the largest PV systems operational in China are the 20 MW facilities at Shenyang and Xuzhou City in Jiangsu Province, which were completed in 2010. Constructed by Chinese polysilicon PV manufacturer GCL-Poly with a reported investment of ¥420 million ($61.5 million), the Xuzhou plant covers 115 acres and once fully commissioned is expected to produce around 26 GWh per year, giving an annual capacity factor of around 14 per cent.[33]

The Xuzhou City plant is just the latest in a small string of large-scale PV projects that have been constructed in China since 2009. Several other 10 MW plants have also been completed, and are listed in Table 5.7.

Four of these plants have so far been constructed in what could broadly be termed the desert regions of China (those in Shizuishan, Ninghia Hui Autonomous Prefecture, Dunhuang City and Younyu). Given that China has plans for up to 13 GW of large-scale solar from its desert regions by 2015, these plants are likely to be just the start of a far larger deployment (see Chapter 6).

Table 5.7 Large-scale PV plants in China (March 2011)

Size	Location	Status
20 MW	Shenyang, Jiangsu	Operational (2010)
20 MW	Xuzhou City, Jiangsu	Operational (2010)
10 MW	Younyu, Shanxi	Operational (2010)
10 MW	Lhasa, Tibet	Operational (2010)
10 MW	Shilin, Kunming	Operational (2010)
10 MW	Dunhuang City, Gansu	Operational (2010)
10 MW	Dongtai, Jiangsu	Operational (2009)
10 MW	Shizuishan, Gobi Desert	Operational (2009)
10 MW	Ningxia Hui Autonomous Region, Gobi Desert	Operational (2009)

Source: www.pvresources.comww.pvresources.com

Large-scale PV in other countries

Outside of the regions already mentioned, there are only three countries with PV installations greater than 10 MW up and running: South Korea, the United Arab Emirates and Canada. Perhaps the most surprising of these is Canada. With its relatively high latitude and long winters, it does not seem the most obvious choice for large-scale solar, yet it is home to several multi-megawatt PV plants, including two 23.4 MW installations in Ontario, at Sarnia and Arnprior. Furthermore at the time of writing an additional 60 MW had just been completed at the Sarnia site, making it the largest PV generation plant. Korea's largest PV plant is the 24 MW Sinan plant, constructed by Conergy, followed by a 20 MW facility near the capital Seoul, a 15 MW plant at Gochang and a 13.7 MW plant at Taean.

For the purposes of this study, however, the most interesting of the large-scale PV plants outside the main markets of Europe, China and North America is the 10 MW generator at Masdar in the United Arab Emirates. Part of the ambitious Masdar Eco-City initiative in Abu Dhabi, this is the first multi-megawatt PV installation in the Gulf States and will provide a useful prototype for the region. Constructed by Environment Power System, the plant uses a mixture of solar cells – half crystalline silicon from Suntech and half thin-film CdTe from First Solar. Built at a cost of around $50 million and covering 55 acres, the plant is expected to have an output of 17.5 MWh once fully operational, giving an impressive capacity factor of nearly 20 per cent. With a cost of $5/W installed capacity, this could make it one of the most cost-efficient power plants currently constructed (see Chapter 6 for more on this plant and the Masdar Initiative).

Conclusion

I hope that by now it will be clear that solar energies, both CSP and PV, are genuine technologies of scale, and ones with wide application that are developing fast. In fact, things are developing so fast that I will be surprised if the details in this chapter are not horribly out of date within a year or two of going to print. I will be disappointed indeed if the idea of an 84 MW PV facility being the largest in the world does not seem quaint in the extreme within five or ten years. That is okay though. The purpose of this chapter is simply to provide a good summary of where things are at the beginning of 2011, but also to show how far things have come in the last five years.

Chapter 6

Long-term visions

I hope that by now it will be clear that large-scale solar energy from the deserts is a real and growing possibility. I also hope that the general reader will have some idea of the technologies that are involved as well as the scale and scope of the projects currently under construction and those that are already in the pipeline, and also a feel for the fluid and fast-moving nature of the sector. Yet as exciting as the current situation is, solar really is just an industry that is beginning to emerge. If solar power of any kind is to play a significant role in helping to tackle the challenges of climate change and fossil fuel dependency it will need to expand by several orders of magnitude. For example, all of the CSP projects currently under construction, or under serious development, will amount to less than 20 GW of electrical capacity, able to supply less than 0.2 per cent of global demand. PV is a little more advanced, installing 7–8 GW in 2009 (likely to be nearly double that in 2010),[1] but this is largely restricted to just a handful of core markets in Europe, Japan and North America. To take the next step, hundreds of GW will be required, and by the middle of the century, thousands of GW. But far from being an impossible task, schemes on this scale are already being discussed in earnest in the corridors of Europe, the Middle East and beyond. Of particular interest to this study are those projects that centre on the world's deserts or semi-desert regions. In Europe and North Africa, grand schemes like DESERTEC and the European Plan Solar de Mediterranean are conducting detailed studies to show what might be feasible, and are working to provide a roadmap. In the Persian Gulf, efforts are being made to construct an innovative new city using a mix of energy efficiency, high-tech architecture and renewable energy. Known as Masdar City, the project plans to offer an exciting glimpse of what is possible – albeit at a high price. While it is very much a flagship project of the Abu Dhabi government, there is some potential for replication in other countries and regions around the world, particularly in developing countries that will need to provide new housing and infrastructure for millions of people in the coming decades. Elsewhere in the world, from the emerging giants of India and China to the copper fields of South America, a vision of a potential solar future is beginning to take shape. The recently launched National Solar Missions of India could see 20 GW of solar electricity installed by 2020, with figures of 200,000 GW proposed by 2050. At the same time some 35,000 square kilometres of the Thar Desert have reportedly been earmarked for possible development. Yet none of these schemes will be simple, nor are any certain. Finance will be scarce, as will the money needed to support and develop the technology in the short term. Land, water and environmental impacts will also begin to come under increasing scrutiny as these projects start to look less like eco-dreams and more like a cold industrial reality. Finally there remain questions as to just how ambitious governments are really being, and if the political will is there to support the programmes behind the rhetoric. This is perhaps the most important question of all. We know that solar technology works, and will get increasingly efficient and competitive. We know that the energy source is adequate to cover our needs, or make a major contribution to our energy mix. We know that we cannot keep burning fossil fuels. Sadly we also know just how hard it can be to fight against apathy and vested interest.

The sections below will look at some of the grand plans being made for harnessing power in the world's deserts. Some such as Masdar rely primarily on decentralised energy, energy efficiency and medium-scale systems (in theory at least, although as we shall see this is changing as the project progresses). Others such as DESERTEC, while I am sure placing equal emphasis on efficiency of use, are nevertheless seeking to provide electricity on a scale that we are currently receiving from fossil fuels. The third is simply a national strategy and target. With the exception of Masdar, where something at least is definitely happening, these are not in themselves projects or 'concrete' plans, but they are still worth considering to provide a broader picture of what is under consideration. From a developer or supplier's perspective, it is also worth remembering that many of these projects are envisaged to come to fruition at around the same time, in the decade 2020–2030. That begins in less than ten years.

As usual, this chapter is not intended to be an exhaustive run-down of each and every plan for large-scale solar, but simply to provide a few examples of what might happen. It is also worth remembering that some of the most promising regions for solar energy from the desert, such as the South-Western United States, are not in the habit of launching grand centralised plans, but are nonetheless likely to see huge development.

Masdar City in Abu Dhabi

Over the last couple of decades, the small Gulf States of the Middle East have become a byword for wealth and opulence. The proliferation of glittering skyscrapers, artificial islands, private jets and ubiquitous four-by-fours are symbols of both conspicuous consumption and the world's insatiable demand for oil and gas that has led to an almost unprecedented accumulation of wealth in the sparsely populated region. Even with the battering they received from the recent global financial crisis, these countries remain among the richest in the world, with sovereign wealth funds controlling trillions of dollars between them. Unfortunately, this explosion in high-consumption living (both among the locals and the wealthy expatriates who have been attracted with promises of low taxation, cheap imported labour and easy money) has led to a growing human and environmental footprint. The per capita carbon emissions of the region are amongst the highest in the world, over 70 tonnes of CO_2 per person in Qatar, and over 40 tonnes per person in the United Arab Emirates.[2] At the same time, the enormous amount of construction going on in the region (at one point it was rumoured that a quarter of the world's cranes were in use in the UAE) has hoovered up precious resources, and damaged the delicate marine environment of the Gulf as millions of tonnes of sand and gravel have been extracted and dumped in the sea. The cities too seem unsuitable for the harsh environment they find themselves in. Walking outdoors for any length of time in summer is almost impossible, with temperatures regularly in the high forties, and the wide roads, steel and glass providing little shade. Compared with the shady alleyways and boulevards of old Cairo or Damascus, or even the reconstructed 'souk' of Doha, these cities seem poorly designed, almost immature.

Yet amidst the ever-expanding skylines and motorways of the region small steps are being taken to envisage a more sustainable future away from oil and gas, and in keeping with recent Gulf tradition, they are grand plans indeed.

Formed in 2006, the Masdar Initiative (meaning 'source' in Arabic) is a state-owned energy business in Abu Dhabi, the richest of the United Arab Emirates. Its professed aim is to turn Abu Dhabi into a world leader in renewable energy technology, helping to diversify the emirate's economy away from oil and gas and attract high-tech industry. As well as investing in traditional energy infrastructure, the company has been heavily involved in a number of renewable energy projects, including the vast 1000 MW London Array offshore wind farm in the United Kingdom and the Terrasol solar power project in Spain.[3] By far its most ambitious project to date though is the development of Masdar City, an entirely new urban development, set in the desert some 17 km from Abu Dhabi city (Figure 6.1).

Covering 6 square kilometres,[4] and with an estimated price tag of $22 billion, the city was originally envisaged as a new model for the twenty-first century, a zero-carbon, high-tech hub for new businesses. At its centre will be the new Masdar Institute for Technology and Research, a joint venture with the Massachusetts Institute of Technology, which is providing advice and helping to form a curriculum and faculty. Aside from the Institute, the city could provide space for up to 1500 businesses including shops, research and manufacturing facilities, along with living space for up to 50,000 people.[5, 6] While the amount of power needed for such a scheme could probably be supplied by just one of the large CSP plants currently seeking planning

Long-term visions 123

Figure 6.1 Artist's impression of aerial view of proposed master plan of Masdar City (southern orientation) (source: Masdar City).

permission in the USA, what sets Masdar apart is the example it intends to set for what a modern city in the Gulf should look like.

Designed by London-based architects Foster and Partners (the firm responsible for dozens of high profile buildings ranging from the renovated Reichstag in Berlin to the United States' first commercial Spaceport in New Mexico), and unveiled in 2007, the masterplan for Masdar City paints a picture of something entirely new: an experimental community in the desert, drawing inspiration from traditional Arabic architecture combined with cutting-edge technology. Renewable energy, energy efficiency and clever design would be employed to create a low carbon, zero-waste space for living and working. The buildings and streets will be orientated to maximise the cooling night breezes that flow from the desert into the sea, while at the same time minimising exposure to the burning desert sun. Large parks, green spaces and tree-lined streets would help to cool the city and make it more pleasant to move around, providing a modern version of the desert oasis. Wind towers, similar to those found in cities like Yazd in Iran, would help to circulate cooler air, reducing the need for energy-intensive air conditioning. Water use will be reduced through recycling and low-water appliances, while waste will be recycled or used to generate electricity and heat.[7]

Solar energy will play its part, with over 100 MW photovoltaic systems planned (the first 10 MW has already been completed and is providing power for construction) and decentralised PV spread across the rooftops of the homes and institutions. Concentrating solar thermal too has been included in the mix, with a 100 MW parabolic trough system in the pipeline (Shams 1).[8] The world's largest hydrogen plant will help to store the energy produced by these systems, while fresh water will be provided by renewably-powered desalination plants.

Even transport has been included in the master plan. Under initial proposals the city would be raised by 7 metres to allow a planned underground train system and providing space for electric cars and pods to move out of sight. The initial version of the plan foresaw cars being kept out entirely, with vehicles being left at the entrance after which passengers would transfer underground to smaller electrically powered 'pods' to take them to their final destination. This design also had the advantage of allowing the town to have narrow streets, providing shade from the fierce desert sun.

Reality dawns?

Put bluntly, the ambitions for Masdar are wonderful. Yet, as might be expected for a project of this scale, things do not seem to have gone quite according to the master plan.

To begin with, the project is running behind schedule. Originally intended for completion in 2016, with the first elements due in 2010, the timetable has begun to slip with recent reports placing the completion at nearer 2020 or beyond.[9] The Masdar Institute, which as the heart of the city was intended to be the first section completed, is behind schedule too, prompting its first Director to resign in 2010, after just one year in charge. While it is now almost complete, those who have visited report that it currently stands alone in an almost deserted city.

There have been problems too with the technology, with the developers pointing to difficulties in getting the renewable components in place. One of the key suppliers of thin-film photovoltaic panels for the city, Applied Solar, has reportedly experienced financial difficulties caused by the recent collapse in the price of poly-crystalline silicon, rendering its amorphous-silicon panels less competitive. This has forced Masdar to look elsewhere for PV.[10,11]

The electric-pod system too has faced difficulties, with many suggesting that it is just not viable and far too expensive. Newer plans for the city see cars allowed, although at present the PR for the developers is insisting that these will still be restricted to electric cars.[12]

Part of the problem for Masdar City is that it is not intended to be simply a vanity project (although there is an air of showmanship about it), but a genuine commercially viable development. Those behind it intend to make money by selling and renting the real estate. Because of this the project has to walk a fine line between maintaining its status as a zero-carbon or near-zero-carbon project and becoming a fudged compromise that may end up pleasing nobody.

Having said that, Masdar is an attempt to do something quite new, and many of its critics may be over-harsh. Very few projects of this scale come in exactly on time and on budget, and the skills and ideas learned at Masdar will no doubt find themselves quickly applied elsewhere.

Masdar – mark II

So what is happening on the ground, and what will the final city look like? The simple answer is that it may be too soon to tell, but perhaps we can make an educated guess. According to insiders, the project is still well under way, but several years behind schedule. It is also likely that the ambition of the original plans will be heavily toned down. Rather than incorporating revolutionary design into the fabric of the entire city, it seems likely that it will contain a mix of new low-carbon, and more traditional commercial architecture, with fewer buildings, but larger and taller, increasing available floor space. There will be a number of showcase buildings, such as the Masdar Institute, surrounded by a host of more conventional structures.

The overall size of the city will also probably decrease, at least in the first instance. In terms of energy, there will be a greater emphasis on large-scale renewable energy generated off-site than on comprehensive energy efficiency measures and decentralised renewables. Already, the first 10 MW photovoltaic plant (half crystalline silicon and half thin-film PV) is up and running, providing some of the energy for construction. As mentioned above, there are currently plans to supplement this with a 100 MW parabolic trough CSP system, and another 100 MW PV plant, again located outside the city. This will take the place of much of the roof-top PV that was originally proposed. In essence it seems that Masdar is gradually morphing into a general city that just happens to be powered largely by renewable energy, rather than something truly revolutionary in which renewable energy and efficiency is the central component.

While 'scaling back' is perhaps to be expected, the biggest failure of Masdar will be if it fails to address crucial issues like transportation and water supply. The idea of conducting motoring within the city in electrically powered pods fed by hydrogen fuel cells is almost certainly dead, and there are doubts about the underground train connection to Abu Dhabi. Without a concerted plan for dealing with road traffic, however, the city will be able to lay only limited claim to the idea of being an 'eco-city'. Road transportation accounts for almost half of all greenhouse gas emissions in many developed countries and is one of several 'elephants in the room' in tackling climate change (others in my opinion being land-use change, population growth and over-consumption of meat).

Similarly water consumption and supply will need to be addressed. The Gulf States lie in one of the most water-stressed regions of the world, with fresh water availability of less than 100 m^3 per person per day. Given the region's rapid population growth, and increasing demands from industry and luxury lifestyles, this has soared. At present the water shortage is met through imports and desalination, the environmental consequences of which have already been alluded to in previous chapters.

Whatever the future of the Masdar city as a whole, the development of solar power in Abu Dhabi and the Emirates looks to be a 'no brainer'. With the region's outstanding solar resources and growing population, it seems like a perfect match. However, the reliance on patronage from the royal families and the lack of a stable support mechanism to encourage solar or guarantee grid access could represent significant stumbling blocks, as could the inescapable contradiction of countries whose wealth relies on oil and gas putting their own efforts into solar power.

DESERTEC and the Plan Solar de Mediterranean

Perhaps the most exciting and advanced vision for large-scale solar that currently exists is put forward by DESERTEC, an organisation dedicated to spreading the vision of large-scale renewable energy in Europe, and from the deserts of North Africa and the Mediterranean region. In a series of scenarios it sets out a detailed roadmap not only to show that renewable energy is the best and cheapest option to meet the energy needs of countries in the region, but also how imported solar from North Africa and the Middle East could be providing some 15–20 per cent of Europe's electricity by 2050.

The Plan Solar de Mediterranean, for its part, appears to be a French-backed version of DESERTEC, with a more modest mid-term plan for solar energy development and transportation. At the time of writing details were still scarce, but it is discussed briefly here.

Background

DESERTEC, as a structured concept, began in 2003 when its founder Dr Gerhardt Knies began to establish a loose network of scientists to explore the potential of bringing together the vast solar and renewable potential of the MENA and Mediterranean regions with new and modern means of high-voltage transmission to provide energy for Europe. Originally known as the

Trans-Mediterranean Renewable Energy Cooperation (TREC), the scientists sought to explore the viability of such a scheme, examining the various options to overcome security of supply, climate change and water shortages.

Despite some purported initial scepticism, the group eventually managed to persuade the respected German Aerospace Research Centre (DLR) to conduct a feasibility study into the issue and in 2005 DLR produced the Med-CSP report, in which it showed that the renewable energy generation and transmission potential in the Mediterranean and MENA region was not only technically sufficient to enable a significant reduction in fossil fuel use, but also financially attractive. Two more reports in 2006 and 2007 looked more closely at the potential for electricity transmission between North Africa and Europe and throughout the region (Trans-CSP), and also at the potential for using renewable energy (particularly waste heat from solar installations) for desalination (Aqua-CSP). Taken together, these three studies probably represent the most authoritative and revolutionary scenario for large-scale renewables that has so far been articulated.

On the back of these studies, TREC was renamed the DESERTEC Foundation and launched as a charity in Germany, with a small staff and a board of advisers. Shortly afterwards the Foundation began to work with Munich RE, a major insurance company which saw the long-term threat that climate change posed to its business and which was keen to explore alternative energy sources. Munich RE then began to gather a collection of major corporate players, and in 2009 the DESERTEC Industrial Initiative (DII) was launched as a commercial enterprise, made up of twelve major companies (this has since grown to 17, with a further 17 associates).

Since 2009, DII has been incorporated as a German company with Paul Van Son as its Chief Executive Officer. While DII pursues commercial angles to promoting renewable energy in the region, and provides the industrial back-up, the DESERTEC Foundation continues in its role as a 'cheerleader' for the idea, raising awareness, conducting research and building support for the concept.[13]

The DLR scenarios

The basic aim of the DESERTEC project is straightforward but incredibly difficult – to find a solution to the multiple pressures of climate change, energy security, economic growth and water availability in the European and MENA region, based on a range of studies produced by DLR. These give a range of visions for the future, depending on technology employed and necessary greenhouse gas emissions reductions. While a look at the reports suggests they have clearly come out in favour of CSP as the main source of energy to achieve this, the concept is relatively technology neutral and the studies conducted by DLR examine the potential not only of concentrating solar, but also of wind and wave power, geothermal, biomass and photovoltaics. They also provide projections on a whole range of issues such as water use, population growth and electricity consumption which underpin the renewable energy scenarios. The current role of both the DESERTEC Foundation and the Initiative is now to encourage the implementation of studies and increase the use of large-scale renewable energy in North Africa and the Middle East, both for domestic use and for sale to Europe. It is not clear if there will ever be an actual physical project called DESERTEC, or if its aims will simply be put in place piecemeal by the companies of the consortium along with other partners, but the latter seems more likely.

At the core of the DESERTEC proposals are the scenarios and feasibility studies conducted by DLR. These lay out in stark terms the problems we will face if the development of the European and MENA regions continues on a business-as-usual model. They then go on to examine various options for overcoming these problems. In every case widespread adoption of renewable energy, and particularly solar, is identified as not only the most environmentally sound solution, but also the most economically feasible.

Climate change and carbon emissions are the first issue to be tackled. Under the current recommendations from the Inter-governmental Panel on Climate Change (considered by many to be too weak), global emissions must fall by around 30 per cent by 2050. In order to achieve this, developing countries will need to cut their emissions by around 80 per cent by the middle of the century, while developing and transition economies will be able to continue increasing their emissions by up to 30 per cent. The general aim is for global per capita emissions of around 1.5 tonnes of carbon dioxide equivalent. Currently, we are a very long way from this. Europe emits around 1887 million tonnes of CO_2,[14] and while this is falling thanks to the economic crisis and some progress in de-carbonisation, it is likely to continue rising slowly in the coming years under business-as-usual scenarios. In the MENA region, the growth in CO_2 emissions is likely to be dramatic. According to the MED-CSP report, in 2000 the Mediterranean countries (European and MENA) emitted around 770 million tonnes of carbon dioxide per year, with estimate suggesting that this will rise to 2000 million tonnes by 2050, a proposition it considers to be unacceptable.[15] At the same time the study accepts that energy and water use will need to increase in absolute terms in order to provide for the increased standards of living in the MENA region, and its growing population.

In order to achieve the necessary level of carbon dioxide emissions, something drastic needs to happen. This is the scenario presented in MED-CSP, which studies the feasibility and necessity of a large-scale switch to renewable energy by 2050. Combining the MENA countries along with the European nations of Portugal, Spain, Italy, Greece, Malta, Cyprus and Turkey, the report looks at what will happen under a business-as-usual scenario, and then looks at what needs to be done to reduce carbon emissions to an acceptable level. By 2050, under the low carbon scenario, fossil fuels in the MENA region will only be used for peaking power, while renewable energy will be able to supply all additional electricity needs. By 2050, conventional power from fossil and

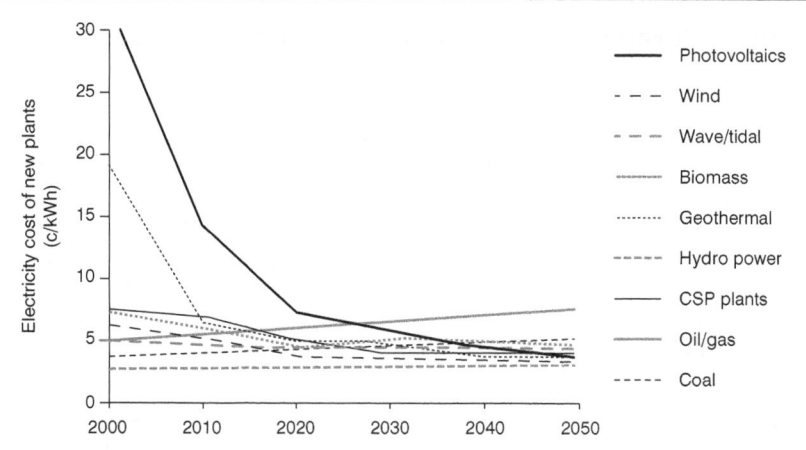

Figure 6.2 Predicted costs of renewables under the MED-CSP scenario (at 2000 prices) (source: German Aerospace Centre (DLR), Institute of Technical Thermodynamics Section Systems Analysis and Technology Assessment, MED-CSP: *Concentrating Solar Power for the Mediterranean Region*.

nuclear fuels will have declined dramatically in importance, from supplying around 90 per cent of the study area's electricity today, to less than 20 per cent by 2050. At the same time there will be a massive expansion of CSP and other renewables, which combined will supply the remaining 80 per cent of electricity required in the region (which at 4200 TWh will be more than three times its current level). As well as being technically feasible, the report indicates that renewables will also be the cheapest way to provide power to the region. Assuming a highly conservative reference price of $25 / bbl of oil and $49 / tonne of coal the report estimates that renewables will require about $75 billion in support to bring them to cost-parity with fossil fuels, sometime around 2020 (this is excluding the 'external costs' of fossil fuels, which would presumably make this happen even sooner). From 2020 until 2050, the study estimates that the switch to renewables will actually save economies in the regions around $250 billion compared with business-as-usual scenarios using fossil fuels. Were the average price of oil to increase more rapidly than anticipated throughout this period, then the savings could be even larger. Figure 6.2 shows the predicted costs of renewables in the region under the MED-CSP scenario.[16]

Crucially, under the MED-CSP scenario, emissions of carbon dioxide are able to fall by 40 per cent in the Med-MENA region, while at the same time providing for increased energy consumption in developing economies.

The second of the DLR reports, Trans-CSP, builds on the work done in the first report, but instead looks at how electricity consumption in the European Union (as opposed to just the Mediterranean region) can be decarbonised. Contrary to what many have written, the report does not simply set out ways for Europe to import energy from North Africa, but also examines the renewable energy which will be available within Europe itself, including wind, hydropower, biomass and geothermal (indeed it estimates around 1520 TWh potential from wind in Europe – around 30 per cent of projected 2050 demand).[17] Under this scenario, renewable electricity consumption in Europe as a whole will increase from around 20 per cent in 2000, to around 80 per cent in 2050. Again, fossil fuels will be retained to supply peaking power, or to cover gaps caused by fluctuations in renewable power, but the overall consumption of fossil fuels will be dramatically reduced. Nuclear energy will be phased out, due to its high cost and the fact that it is only economic if it is running at a constant output, something that DLR believes makes it incompatible with a renewable-orientated energy mix. As in the Med-CSP studies, the calculations estimate that focusing on renewables now is the cheapest long-term solution. If timely investment is made in renewable energy, the report argues, it could be cheaper than fossil fuels by 2021 (excluding as always the external costs of fossil fuels).[18]

The second major aim of the Trans-CSP report is to look at the potential for renewable electricity generated from CSP in the MENA region to be transmitted to Europe in order to complement European renewable energy sources. This is an important part of the concept, as renewable energy generated from CSP sources in MENA countries can provide reliable baseload electricity, and so help to take the place of sources like coal and nuclear that the scenario sees being phased out. To begin with, the report looks at the various options for transporting energy, including conversion to hydrogen, transport through the AC grid system, or use of an HVDC transmission system. Both hydrogen and AC transmission are discounted thanks to their predicted energy losses (75 per cent and 45 per cent respectively) in the conversion and transportation process (estimated to be up to 3000 km). This leaves HVDC as an option. In its study, DLR estimates that a well-designed HVDC system could have transmission and conversion losses of around 10–15 per cent over 3000 km, allowing transportation from the MENA region to major load centres in Europe. They also suggest that HVDC can be more environmentally friendly than standard AC transmission since it requires fewer power lines and hence a smaller land-area.[19]

Starting with a potential energy transfer of 60 TWh in 2020–2025 transmitted through two 5 GW HVDC cables, Trans-CSP outlines the potential for the MENA region to supply 700 TWh of solar electricity to Europe, at a cost of €0.05 / kWh, including transmission losses. This would require a total of twenty 5 GW transmission lines, powered by 2500 km² of CSP installations. Combined, these would be sufficient to supply around 15 per cent of Europe's projected electricity needs in 2050, and generate jobs and income for the MENA countries.

Aside from simply outlining the technical feasibility and necessity of using renewable energy for electricity, another interesting feature of the DLR study and the DESERTEC plan is outlining the possibility of using CSP for large-scale water desalination. The countries of North Africa and the Middle East are some of the driest and most water-poor regions on earth. Currently only four – Iran, Iraq, Syria and Lebanon – have renewable water resources above the general threshold for water stress or poverty, generally set at 1000 m³ available per person per year, and these resources are declining fast thanks to economic and population growth. By 2005 the region as whole already had a demand of 270 billion m³ and a water deficit of around 50 billion m³ per year, and this is set to grow. Under business-as-usual scenarios water demand might be expected to reach 570 billion m³. However, this has been rejected by DLR and others as highly unrealistic as that much water will not be available to the region with its current climate and the potential technology. Under lower growth scenarios, water consumption is expected to reach 460 billion m³ by 2050, resulting in a deficit of around 150 billion m³ per year – more than twice the volume of the Nile. Even a scenario involving extreme water efficiency would only reduce this total demand to 390 billion m³, leaving a deficit of just over 100 billion m³. Currently the region's water deficit is met through the exploitation of non-renewable water reserves, such as underground aquifers, and to a lesser extent by fossil-driven desalination. In the future neither of these options will be feasible. Over-exploitation of ground water is already occurring, leading to the drying out of oases and a falling water table, and cannot be sustained or expanded indefinitely. Similarly, most long-term water plans have ignored large-scale desalination due to its high energy requirements, costs and environmentally destructive nature. This is where DLR believes that CSP could help.[20]

As outlined in the technology section, CSP can power desalination in two ways – either by providing the necessary electricity or by providing heat directly to the desalination process. In fact by coupling desalination and CSP plants it may be possible to supply CSP's need for cooling water too. According to the Aqua-CSP study, by 2025 it is likely that CSP will become the lowest cost source of water (< €0.4/m³) and electricity (€0.04 / kWh) in the MENA region, and from this point onwards will be able to replace the over-extraction of ground water. Greenhouse gas emissions from desalination will fall, peaking in around 2025, before CSP technology replaces conventional fossil-driven plants. In order for this to happen, however, it is necessary for the countries in the region to begin the process of developing and deploying this technology for electricity and other uses, and this is essential to lowering the price.

Implementing DESERTEC's plans

The next step for DESERTEC is to build on the work done by DLR, and in particular on the construction of renewable energy in the MENA region, both for domestic purposes and for transmission to Europe. According to the DESERTEC Foundation, this involves two things – raising awareness of the potential opportunity that is there and that is outlined in the DLR reports, and the direct involvement of the DII partners in constructing, financing and developing renewable energy infrastructure in the MENA.

But what would developing renewable energy on that scale actually involve? Well, as mentioned above, supplying 15 per cent of European Union electricity is a big deal, involving at least twenty 5 GW HVDC transmission lines and a huge amount of renewable energy production capacity. 100 GW of parabolic trough systems would equate to about 2500 km^2 of CSP installations constructed by 2050, necessitating a total investment of some €350 billion. Combined these would produce around 700 TWh of electricity. Assuming that the CSP-for-export was spread over four countries (perhaps Morocco, Algeria, Egypt and Tunisia) this would necessitate roughly 625 km^2 of CSP in each country (25 km × 25 km), with five transmission lines heading to Europe. These lines would require an additional 3600 km^2 of land between the four nations (although it would be a narrow strip), and another €45 billion in investment.[21]

So how does the project envisage getting there? According to the timeline outlined in the Trans-CSP report, by 2020 Europe should ideally be importing around 60 TWh of renewable solar electricity. This would require the construction of 10 GW of CSP spread over 225 km^2 and two new transmission lines in the next ten years, with a total investment of €47 billion. Currently, there are only around 1–2 GW of solar energy under development in the MENA region.

Meeting the needs of the domestic markets in the MENA countries will also need to be conducted in parallel, and this could require many more gigawatts of CSP and PV by 2020. Producing 2000 TWh for domestic consumption would necessitate roughly 400 GW CSP, spread over about 10,000 square kilometres of the MENA region.

Next steps

According to DESERTEC, the idea is to continue planning and preparations for the next three years, with part of this process examining the feasibility of expanding the DESERTEC concept to other parts of the world, such as India. Following this the consortium hopes to begin construction of its initial plants, with the aim to have the first generators on-line by around 2015.

Who will pay for it?

Obviously one of the first questions that will come to mind to anyone reading this might be, who will pay for such grand schemes? Figures like €395 billion to produce and transport electricity to Europe seem prohibitive, but it is important to put statistics like this into context. Huge investment will be required in the coming decades in any case simply to maintain and replace Europe's existing energy supply, regardless of the energy source. Furthermore, since it is estimated that shortly after 2020, renewable energy sources will become among the cheapest available, this is actually less than it would cost to supply a similar proportion of Europe's energy needs from conventional generation. This is certainly the finding of the DLR, which estimates that supplying an equivalent proportion of Europe's electricity supply from coal or natural gas by 2050 would cost significantly more, and if the costs of the environmental destruction were included, spectacularly more.[22] The trouble, as ever for renewables, will lie in the short term – how to develop enough momentum to allow the industry to keep growing and reduce costs. The Med-CSP report estimates that it will take the investment of around $75 billion by 2020/2025 to enable renewable energy technologies like CSP to reach a point at which they 'break even' with fossil fuels. After this they will become the cheapest source of electricity, saving a further €150 billion to 2050.

For Europe, the DLR report is clearly supportive of feed-in tariffs like those found in Germany and Spain (while the Spanish PV sector may have suffered thanks to an ill conceived FIT, it is worth remembering that a similar programme has allowed it to build a world-leading wind industry). Feed-in tariffs, the report argues, provided they are reduced steadily over time, and ultimately phased out, do not need to be considered as subsidies but rather as public investments. The report also argues, as has been stated elsewhere, the need to reduce funding and subsidies to the fossil fuel industry, which will help level the playing field.

For developments in the MENA region the picture is more complex. Some, such as Algeria and Morocco, may consider introducing their own feed-in tariffs, which should help to spur growth. Electricity that is directly exported to Europe may benefit from the generation of CDM credits under the Kyoto Protocol (or whatever its successor will be called). Almost all of the countries in the region have ratified the Kyoto Protocol and so could be able to take advantage of schemes like the CDM. Other renewable projects may be built directly with government backed investment from Europe in exchange for guarantee of renewable power supply. In the longer term, as the costs of solar fall government investment needs should be minimal, with private companies and individuals able to spur greater developments.

Criticisms and problems associated with DESERTEC

Aside from the obvious technical, economic and political challenges that must be overcome to make DESERTEC's vision a reality (basically persuading government and business to invest the capital, and persuading politicians and the public that it is worth doing), a number of other concerns have been raised about the plans.

Perhaps the thorniest issue is over control of the solar generators and resources. This will be a difficult problem to solve. European companies and governments that have invested billions of euros in developing renewable energy generation capacity will need assurances that their energy supply and investment will be protected, and that they will not be held to ransom essentially swapping one form of energy insecurity for another. Russia's disputes over gas supplies with its neighbours have shaken Europe and shown what a powerful position an energy exporter can be in. At the same time, Middle Eastern countries need to be sure that they too will benefit and some commentators have voiced fears of resources colonialism, and fear that they will benefit little from the electricity they generate and will become offshore satellites of European nations. While it is too soon to give a definitive answer to these questions, simply looking at the DESERTEC plan shows that the extremes of either scenario are unlikely. Firstly the system will be decentralised across many countries, and in the end there will be up to twenty HVDC lines transporting electricity to Europe. This would greatly reduce the likelihood that Europe's power supply could be easily cut – whether by political actions or through terrorism or natural disasters. Certainly the situation would be no worse than Europe's current reliance on natural gas from Russia, Algeria and Norway, and recent turmoil in the region shows how unstable reliance on oil is in any case. Middle Eastern countries too seem likely to benefit through the availability of cheap electricity to meet their growing needs, employment opportunities that the massive solar expansion would bring and the revenues that they would earn from licensing the sites to solar companies. Nothing is guaranteed, however. Too many resource-rich countries have been blighted by an over-reliance on natural resources and have failed to improve the standards of living for their people. The role of poor governance and corruption cannot be overstated and it would be naïve to think that DESERTEC-like projects could not simply lead to a situation of Middle Eastern elites lining their pockets while their people derive little benefit.

There may be political problems too. Relationships between many MENA countries are tense (Algeria and Morocco over Western Sahara for example) and the region as a whole is one of the most unstable parts of the world. Significant cross-border cooperation may be required to allow DESERTEC to thrive and wider political problems pose a significant risk to this. Already there are news reports that Algeria has snubbed the DESERTEC idea, claiming a lack of consultation. While it is easy to feel exasperation over cries of non-consultation for a project that does not yet exist, anyone who has worked in international politics in any form knows just how crucial an issue this can be. If DESERTEC, or similar projects, are to be a success it must be with the full participation of the MENA and European nations from a very early stage. (It may be of course that Algeria is simply throwing its lot in with the French-backed Trans-GREEN projects, preferring a partner it knows, but the point stands.) There is also the issue over whether the current wave of instability in the Middle East will have an impact on the future of DESERTEC (as this book is being finished popular protests have overthrown dictatorships in Egypt, Libya and Tunisia, civil war is threatening in Syria and there are angry rumblings in Yemen, Bahrain, and Saudi Arabia). On a more positive note though, it is equally possible that the changes sweeping the region will prove to be positive to renewable energy. Should more open and democratic institutions be established this could provide a more welcome environment for investment, while the region as a whole may receive a renewed burst of energy and activity.

Finally, aside from the political and technical concerns, there are some physical concerns over DESERTEC and its impact on the local environment. Initial fears over water and land use have probably been overdone, however, particularly given the scale of the region and the potential for CSP to produce fresh water (see above). For more information on the challenges associated with land use see Chapter 7.

DESERTEC II? The Plan Solar de Mediterranean and Trans-Green

Seemingly inspired by the DESERTEC idea, in 2007, President Sarkozy of France launched the Plan Solar de Mediterranean (PSM) at the inaugural meeting of the Union for the Mediterranean in Marseille. The general idea is very similar to DESETEC, and indeed potentially complementary to it, but with a more modest goal of installing 20 GW of renewable energy in North Africa by 2020.[23] Again the emphasis is on concentrating solar thermal technology, although perhaps with a larger space for French business than in the more 'German-led' DESERTEC Industrial Initiative (in general French companies have invested far less in developing solar expertise than their German or Spanish counterparts).

In 2010, the PSM launched the Trans-Green initiative, a consortium of French companies with plans to kick-start the process with two 5 GW transmission cable from Tunisia and Libya to Italy.[24, 25] While this would be a large inter-connector in its own right, it will be just the first of many if the vision of Europe sourcing 15 per cent of its electricity from North Africa is to become a reality. Nonetheless it would be an important step, and would fulfil half of the DLR scenario that envisaged the 5 GW HVDC connectors between Europe and North Africa by 2020. Whether this project will be completed, scrapped or delayed in light of the recent turmoil in Tunisia and Libya, remains to be seen, and could have a lot to do with the attitude of any new governments to the European countries, particularly those such as Italy and France that have recently been very close to the former leaders of Libya and Tunisia respectively.

The emergence of a 'rival' DESERTEC has met with a mixed response. Many, including some of those in the DESERTEC Foundation, welcome it as increasing the momentum behind the project concept and spurring competition.[26] Others worry that it may lead to meaningless political rivalry. There have already been suggestions that the French government may have annoyed the Spanish authorities by announcing Trans-Green during the Spanish Presidency of the European Union, at a time when Spain was trying to get European Union backing for the DESERTEC idea. Only time will tell if there will be political room for both initiatives, regardless of if the companies think there is plenty of work to go around between them. It would be a shame if petty national rivalries led to problems and delays within such an important initiative.

India's National Solar Mission – 200,000 MW by 2050?

In 2008, as the countdown was already beginning to the climate change talks in Copenhagen, the government of India released a National Action Plan for Climate Change (NAPCC) – an eight-part strategy prepared by the Prime Minister's Council on Climate Change. In the document, climate change is recognised as a clear and present danger to the people of India, primarily from the point of view of impeding economic development. As a country where the majority of people are still extremely poor, the NAPCC makes it clear that development is the primary goal for India, and that climate change is a challenge to this goal. Crucially, however, the plan also makes the pledge that India's per capita emissions will never exceed those of developed countries (although given the size of India's population and the growth in pollution it would have to go through to reach a similar level of emissions, this is of limited comfort to those concerned by the climate). It is also unclear at this stage if this cap will be set at the emissions of developed countries today, or in the future, when they will hopefully be far, far lower. Nonetheless, this is an important statement from a country where simply getting by is a day-to-day concern for hundreds of millions.

The Action Plan sets out a pathway to combating and coping with climate change, broken down into eight separate 'National Missions'. The complete list consists of the National Solar Mission, National Mission for Enhanced Energy Efficiency, National Mission on Sustainable Habitat, National Water Mission, National Mission for Sustaining the Himalayan Ecosystem, National Mission for Green India, National Mission for Sustainable Agriculture and the National Mission for Strategic Knowledge for Climate Change. Sadly, many of these initiatives seem less impressive on inspection, and as ever the important part will be in the detail and the implementation. Nonetheless one cannot help but be encouraged by the broad scope of the topics, which at least helps to put the problem in context. There are many rich nations which could learn from the concept of a comprehensive national environmental policy.

Jawaharlal Nehru National Solar Mission

Of the eight national projects, the National Solar Mission and the importance of solar energy are given pride of place. Indeed the preamble to the report contains explicit references to the importance of the sun as a provider of renewable energy, stating: 'Our vision is to make India's economic development energy-efficient. Over a period of time, we must pioneer a graduated shift from economic activity based on fossil fuels to one based on non-fossil fuels and from reliance on non-renewable sources of energy to renewable sources of energy. In this strategy, the sun occupies centre stage, as it should, being literally the original source of all energy.'[27]

Renamed the Jawaharlal Nehru National Solar Mission in 2009 after India's first Prime Minister, it calls for the Republic to harness its exceptional solar resource (around 5.5 kWh / m² per day) and increase the amount of renewable energy, and nuclear energy, in its mix. Requiring an estimated investment of around $19 billion up to 2022 (more on this below), the project would support a wide range of measures including medium temperature domestic solar heating systems, the expansion of rural electrification, and increase in grid-connected solar generation and the development of an Indian solar manufacturing and research base. Although the initial NAPCC sets no concrete targets for solar capacity, subsequent announcements have set the goals for solar energy of 20,000 MW by 2022, rising to an impressive 200,000 MW by 2050. As will be seen, however, these targets are conditional on a great many things, not least external funding and the ability of Indian industry to develop a low-cost, indigenous, solar capability. Furthermore, these figures must be put into context. Although 200,000 MW of solar would almost equal India's current installed capacity for electricity generation – around 200 GW in 2010 – this is expected to grow dramatically by 2050, to nearly 1300 GW under business-as-usual scenarios! By this measure solar would still account for only a small percentage of total electricity needs.[28]

Targets for solar

While the futuristic targets have attracted all of the media attention, it is the interim and mid-term objectives that are of most immediate concern (see Table 6.1), and these are distinctly modest. From 2010 to 2022, the aim is to build a *framework* for the deployment of 20 GW of solar electricity by 2022, with two mid-term goals of 1000 MW by 2013, and 4000–10,000 MW by 2017.[29]

At present, little distinction is made between PV and CSP in the Mission document, and also little indication whether the targets will be met largely with distributed rooftop applications or ground mounted systems. It will likely be a mixture of both. However, plans are included for a number of CSP pilot projects as part of the research and development activities. These are a 50–100 MW plant of undetermined technology with 4–6 hours worth of heat storage; a 100 MW parabolic trough facility; a 100–150 MW solar hybrid plant with coal, gas or biomass; and one or more power tower projects of 20–50 MW with or without storage. According to the Mission document,

Table 6.1 Phases and targets of Jawaharlal Nehru National Solar Mission 2010–2022

Stage number	Application segment	Target for Phase 1 (2010–2013)	Target for Phase 2 (2013–2017)	Target for Phase 3 (2017–2022)
1	Solar collectors	7 million square metres	15 million square metres	20 million square metres
2	Off-grid solar applications	200 MW	1000 MW	2000 MW
3	Utility grid power, including rooftop	1000–2000 MW	4000–1000 MW	20,000 MW

Source: Jawaharlal Nehru National Solar Mission, *Towards Building Solar India*

these plants will be constructed after a consultation and bidding process and will be commissioned in India's twelfth 5-year plan (2013–2018). Given the rapid development of solar technology elsewhere, one suspects that the main purpose of these prototypes is to develop and encourage an Indian manufacturing base for concentrating solar thermal technology. Indeed, by 2015 (the middle of the proposed commissioning period) projects on this scale will appear decidedly commonplace, as events in the Mediterranean region, the USA and potentially China gather pace.

Furthermore, by reviewing the document it can be seen that while the headline targets seem progressive, there are numerous details that could bring the NSM back to earth with a crunch. To be fair, 1000 MW of new grid-connected solar by 2013 is reasonably ambitious, given India's starting point of just 30 MW installed in 2010, but 4000 MW cumulative capacity by 2017 seems a very low objective, particularly given the size of the country and the experience of what can happen elsewhere if the conditions are right (over 3000 MW of PV alone in a single year in Spain or Germany). However, things get even more uncertain with the higher targets. While the 1000 MW and 4000 MW targets could be supported through a mixture of renewable purchase obligation and feed-in tariffs, the official Mission document makes it clear that higher targets are dependent on international technology transfer and financial support. According to an article published in the *Guardian* newspaper, this represents a significant shift from earlier drafts of the Mission, in which it was suggested that the Indian government could be largely responsible for funding the deployment.[30, 31] Recent versions available from the Ministry of Environment and Natural Resources (MNRE) contain no specific figures on investment costs, but numerous references to the UNFCCC funding process. Indeed, prior to the Copenhagen climate talks the Indian Minster for the Environment at the time, Jairam Ramesh, supported calls for the developed countries to provide $200 billion a year in funding for developing countries.[32] Although one can certainly see his argument, given the historical inequalities in carbon emissions, practically speaking it might be hard for the western democracies in particular to justify such transfers of money and technology to an increasingly powerful country such as India, particularly when their electorates will be suffering from austerity measures and attempts to restructure their own economies. Despite their massive flaws, instruments such as the Clean Development Mechanism may prove to be more workable tools, and enable the funding of specific projects, although as yet there is no clear indication of how this will develop in the post-Kyoto phase.

Manufacture and research

Crucially, the solar Mission aims to invest not only in the deployment of solar technology, but also in manufacture, research and development, with the aim of driving down costs, making India a world leader in the various solar technologies. Targets for the industry include a manufacturing capacity of 4–5 GW by 2020, a not unreasonable target given the country's current photovoltaic production capacity of around 700 MW. In fact given that the report estimates the top 15 manufacturers in India already have plans that would see PV production reach 8–10 GW by around 2020,[33] one could argue that this target actually lacks ambition, particularly if India wishes to be a major exporter on the global PV scene. Alongside the manufacture of photovoltaic cells and modules there are also plans to greatly increase the domestic supply of poly-crystalline silicon, almost all of which must be currently imported. Importantly though, the current plans make no recommendations for developing an indigenous capacity for CSP development (aside from the implications in the research installation mentioned above). Since this technology currently offers a strong hope for large-scale utility solar, it can be assumed that unless something changes, India will need to import a significant proportion of its solar technology.

Research too gets attention in the plan, with the ultimate aim of helping the industry develop and driving down costs to achieve grid-parity for solar (i.e. it will be directly competitive with other technologies) by 2020. Steps outlined to achieve this include setting up a Research Council and a National Centre of Excellence to promote solar research and development.

Regulatory changes

In order to facilitate the solar mission, the Department of the Environment has recognised that there will need to be changes to legislation and regulation. In particular it has explicitly stated that it will be necessary to modify the National Tariff Policy 2006 to mandate a fixed percentage of electricity from solar technology in each state. Current thinking puts this at about 0.25 per cent rising to just 3 per cent by 2020. Potential investors in India, however, well aware that what the central government mandates and what the states implement can be quite different, are likely to wait until a definitive policy is implemented before major decisions.

A start

Perhaps the best thing that can be said about the Indian National Solar Mission is that it is a start, and as so often that is the most important thing. Over the last ten years renewable energy has consistently beaten all but the most optimistic estimate of what could be installed and the speed with which manufacturing could be scaled up. If the Indian NSM can get the ball rolling, and provide an appropriate framework there is no reason that this cannot happen again. 200,000 GW by 2050, sounds a lot, but in reality it could be just a fraction of what is actually achieved.

Energy from the Thar Desert?

One of the most obvious areas for long-term solar power development, and the one most keeping with the theme of this book is the Thar Desert. With a land area of around 200,000 km² and an average insolation of around 21 MJ/m²/day it would be perfect for solar energy deployment. Clearly others are thinking the same, and indeed one of the new spin-offs from the DESERTEC idea, DESERTEC-India, has recently set up a CSP-Thar website to promote the possibilities of renewable energy in the region. Despite its potential though it is unclear yet if there are any concrete or well thought-out plans. The Rajasthan government in whose state the majority of the desert lies, has published a draft strategy that includes mention of using areas of the desert for renewable energy production (although rather worryingly it refers to extensive wastelands, something sure to set fans of the desert on edge).[34] A number of news reports put the amount of land set aside as up to 35,000 km² between Rajasthan and Gujarat.[35] If true this would represent a vast area for development, which would likely be sufficient to host the majority of India's 200,000 proposed megawatts. Indeed, according to back-of-the-envelope calculations from DESERTEC-India this would be enough for 1400 GW of concentrating solar.[36] Other reports in the Indian press have linked the William J Clinton Foundation (set up by the former US President) to investments in solar in the region, with talk of 3000–5000 MW of developments.[37] While the potential is certainly there, for now though it is probably best not to get carried away with developments in the region. Whatever plans there are in India are still in the early stages, and there is some way to go before the country builds even a few large solar plants, never mind thousands.

Conclusion

The three examples of large-scale visions outlined above (and they are just three examples) provide very different views on what is happening in solar energy from the Desert. Masdar is an attempt to create an innovative new city in the heart of the desert, marrying decentralised power, architecture, efficient transport and urban planning. DESERTEC is an industry- and NGO-led attempt to show the sheer potential available in a region to combat environmental, resource and water challenges, and supported by a number of well-researched scenarios demonstrating what can be achieved. The real challenge for DESERTEC will be in bringing the necessary players and political will together to achieve progress in a volatile region. Finally, the Indian programme is a good example of a national strategy, produced with the aim of highlighting a country as a potential leader in this exciting new field. Many countries are producing statements and documents along the lines of India's National Solar Mission, but the challenge for all of them will be turning government desire into affordable and effective policies to encourage the growth of solar energy and reap the environmental and economic benefits it can bring.

Chapter 7

Environmental and resource issues facing solar technology

Solar power, both CSP and PV, has the potential to deliver energy on vast scales, through decentralised rooftop applications and large desert-based arrays. Growth in this industry is rapid across all fronts, and the plans that have so far been reviewed are ambitious and far reaching. If all the projects currently in development are completed, an additional 9 GW of CSP will be constructed in the next five years, more than tripling the current installed capacity. If PV continues to grow at its current rate, annual installations could hit around 35 GW by 2014,[1] and perhaps more than 100 GW a year by 2020. And if the long-term visions outlined in this book come to pass, by the middle of the century, there will be more than 500,000 MW of solar capacity in North Africa and India alone. Yet the future of this industry is by no means secured. There are a number of barriers and complications – environmental, resource, economic, and cultural – that will need to be overcome to ensure that this technology really can make a difference to overcoming the environmental and supply challenges faced by human society. Several of the economic issues have been dealt with elsewhere but this chapter will take a short look at some of the environmental and resource issues.

Environmental and health challenges of solar power

One of the most crucial things to consider when any new technology is being developed at scale is its environmental impact. Although solar is still a relatively small-scale endeavour it is important to look at these issues early on, both to highlight the limitations of a technology and also to look at what measures can be put in place to reduce any negative impacts it might have. This is particularly true of renewable energy technologies, since so much of the justification for their development is environmental.

There are undoubtedly some in the renewable energy business who resent the way that the industry is subjected to the kind of environmental scrutiny that few others must endure, and this is understandable. People rail against the water use of concentrating solar developments, yet that anger is rarely directed against conventional power stations or cut flower producers. Concerns about cadmium from thin-film photovoltaic cells are not matched with equal concern for the heavy metals emitted by coal burning, or contained within each new generation of laptops and televisions. A recent story on a news website reported the concern that aquatic insects may mistake the shimmering surface of crystalline silicon panels for water courses, and attempt to lay their eggs on them in vain. This prompted a flurry of comments on the page deriding the 'so-called' benefits of solar. Only a few questioned how many insects perish due to drying rivers, car headlamps or every time a tree is cut down.

Renewable energy, however, is not just another business. It is part of something much larger, and if done properly will allow us to fundamentally re-order our economy on more environmentally friendly lines. Emerging environmental problems must be tackled directly, not swept under the carpet. This is particularly the case when public support is so crucial to the success of the technology. Yes, the drawbacks of solar must be put in context (and compared with most technologies they are limited indeed), but they should not be ignored.

As an example, the progress of wind power in the UK, and in many other countries, has been held up for years thanks to a potent mixture of misinformation from anti-wind campaigners and some inappropriate commercial wind developments providing them with ample ammunition. Locating wind (or solar!) energy developments in scenic areas or places of national importance for wildlife is a sure way to alienate those most likely to support renewables and incite cries of hypocrisy and greenwash. Similarly, constructing an installation on a blanket peat bog,[2] releasing thousands of tonnes of trapped carbon dioxide provides those who seem to resent the mere existence of renewable energy with yet another example of 'green stupidity' and splits the environmental lobby, pitting wildlife conservationists, developers and climate activists against each other. The fact that many of the largest players in the renewable energy business are the same oil, gas and utilities customers that have shown scant regard for environmental concerns over the decades hardly helps. Similarly, as investment in the renewable energy sector has increased many people have become involved for who the environmental imperative is of little interest. I remember feeling a wave of despair when I saw the workers going to an offshore wind farm near Liverpool casually flinging their cigarette butts off the supply boat into the clear waters of the Irish Sea.

While in some ways the fact that the renewable sector is drawing in a wide range of personnel is to be welcomed, it means that it cannot and will not simply be assumed by the public or the regulators that those behind a project have worked to achieve highest environmental standards, just because it is 'renewable'.

Another good example of how failing to address drawbacks early on can affect an industry is with biofuels, particularly those produced from food or energy crops ('first generation biofuels' or 'agrofuels'). While the general concept of biofuels seems attractive – producing low carbon fuel from 'natural' resources – it quickly became apparent that any attempt to scale up the technology would have serious consequences. With the exception of Brazil and potentially the USA, there are no major countries in the world where biofuels make a serious contribution to transportation. To change this using first generation technologies would require a colossal conversion of farmland to energy crops, with consequences for global food prices. The alternative is to bring new land into agriculture, something with serious negative environmental consequences, not least the release of huge amounts of greenhouse gases. For regions such as Europe or much of Asia, the land is simply not available for biofuels to make a serious contribution to powering our vehicles, and feed-stock would need to be imported, often from countries where deforestation is a major concern. Indeed, far from being green, it has become clear to most that biofuels produced from deforested land are probably worse for the climate than just continuing with burning oil, with the additional devastating consequences that come from forest loss (such as loss of biodiversity, desertification, flooding, forced migration, changing weather patterns etc.). Similarly converting food crops into fuel has potential consequences for global food security. In 2007 high food prices sparked riots across the developing world and while biofuels currently play only a minor role in global food prices, the consequences of their large-scale expansion are obvious.

Sadly, when these arguments were first made, many in the industry simply did not want to hear and pushed governments hard to adopt biofuel mandates. In 2005, I remember attending a All-Party Parliamentary Group meeting on biofuels at the Palace of Westminster in London at which I was publicly shouted at by two gentlemen (one from a farmers union, the other from a biofuels company) for suggesting that there were serious environmental issues to consider before the UK Government introduced a mandate for biodiesel or bioethanol. This was their seemingly green 'golden egg' and they did not want environmental issues to get in the way.

Just a few years later and the limitations of first generation biofuels are plain for all to see. Governments have begun to realise the dangers in bringing in strict mandates, and environmental campaign groups have lined up to say that biofuels must only be used if they have no negative social or environmental consequences. On-going moves by the European Union to increase its biofuel use will likely be controversial and will be fought by environmental campaigners each step of the way. As a result the entire concept of large-scale first generation biofuels has been compromised, and rightly so. The downside of this is that it will probably make the job of those promoting more sustainable sources of biofuels more difficult, since all sources run the risk of being 'tarred with the same brush'. Yet this need not have been the case. More open and honest discussion of the technology's limitation may have led to a much less ambitious, but still environmentally beneficial biofuels industry, potentially supplying niche markets where electrically powered engines are currently impractical (aviation for example). At the same time research could have continued (as it still does) on finding more efficient and environmentally friendly ways of producing biofuels which do not increase the pressure on land or conflict with food crops.

Barriers, concerns and environmental issues with solar power

Moving back to solar power, the main environmental concerns can be divided into several categories. First there are those associated with the production of solar technology. As with other electronics, many types of PV require toxic chemicals in their manufacture, and silicon production is an energy-hungry business. The elements involved in some types of thin-film cells are also rare and finite, and their availability must be factored in.

Then there are issues of water use, particularly in the sort of desert scenarios this book has focused on. By definition water is a precious resource in the desert and there are already concerns that a large-scale expansion of CSP will conflict with water for wildlife, agriculture, and other human uses.

Next, there is the issue of land use. Solar is a decentralised source of energy, and providing large amounts of electricity will require a lot of land. Those figures suggesting that a certain percentage of power can be produced by covering 5 per cent of a desert disguise the fact that we are still talking about big areas, with far-reaching consequences for the landscape, wildlife and people. It is worth investigating these issues, if only to separate genuine problems from misinformed negativity.

Finally, and crucially, there is the question of the life-cycle greenhouse gas emissions of solar photovoltaics and CSP. Are the 'carbon economics' of these technologies rational, and do they produce more power than they use? While for many it seems intuitive that solar panels or concentrating installations will produce far more power over their lifetime than was used in their manufacture, this remains an important area of confusion and a focus of misinformation, with commentators occasionally still claiming that PV can never pay back the energy invested in it.

Health and environmental concerns of PV and CSP manufacturing and disposal

As with any industry, the manufacture and disposal of solar cells and modules carries certain risks and environmental impacts, depending on the type of cell in question. Solar cell manufacture involves the use of a number of toxic and volatile substances, and there is a risk that as systems reach the end of their lives, they will not be disposed of in an appropriate way. None of these problems is unique to solar, however, and none is insurmountable (the flat-screen television and computer manufacturing industries face many of the same challenges).

CSP installations by contrast use standard industrial components: steel, glass, electronic, hydraulics and generators. The safety procedures required for handling this sort of equipment are standard for many industries and there is no need to go in to them here.

Silicon cells

The most common kind of solar cell – crystalline silicon – poses few significant risks to the environment. Silicon itself is relatively harmless unless turned into a fine dust and inhaled and this could only happen if modules were dismantled and the cell itself was ground down, a situation that seems very unlikely. The other main component of a solar panel – glass – is also made largely of silicon and is harmless. However, while there are few dangerous chemicals in a silicon solar panel itself, various dangerous chemicals are used in the manufacture of these cells and could pose a risk if not handled correctly.

For example hydrofluoric acid (HF), nitric acid (HNO_3) and sodium hydroxide (NaOH) are used to clean the silicon wafers and remove oxidised residue. While such chemicals are extremely hazardous on contact or if released into the water supply, their use in industrial applications is routine and safety measures are well established.

Similarly a number of toxic gases are used in the manufacturing process, including hydrogen bromide, chlorine and phosphoric oxide trichloride. Again if released these are harmful, although with proper maintenance of equipment this should be minimal. As an aside it is worth remembering that several of these sorts of halogen compounds are powerful greenhouse gases in their own right, many thousands of times more powerful than carbon dioxide, and their use is a significant contributor to global warming (F-gases account for 1–2 per cent of total greenhouse gas emissions). Thus they must be used as sparingly as possible and properly destroyed. Failure to do so could harm the environment and have an impact on the life-cycle emissions of the solar panels.

Aside from the cells themselves, lead-based solders may also be used in the solar modules to secure the electrical connections, and as with any lead-based application waste minimisation and recycling should be the order of the day, or a switch to different solder materials may be possible. Early PV modules did not comply with international leaching standards and so could not be landfilled. In recent years, however, the regulatory framework has changed and in many cases the landfilling of unprocessed electrical waste is prohibited in any case. Again this necessity is not restricted to PV modules, silicon or otherwise, but to many electronic products. In Europe for example, the recently introduced Waste Electrical and Electronic Equipment (WEEE) Directive made it mandatory for the manufacturers to collect and recycle electronic equipment containing harmful substances. Although solar panels are not currently covered by the WEEE directive, it seems sensible that they should employ the same environmental precautions as other electronics, and indeed several do. To date there are not many end-of-life panels out there, simply because they last for decades and are a relatively young technology, but it is important for manufacturers to factor in the reclamation and recycling architecture in advance of future growth.

PV manufacturing also uses significant quantities of cleaning agents and solvents in cutting, shaping and polishing the solar cells. Standard waste minimisation practices including material recovery and in-house neutralisation of acidic or basic solutions help to minimise hazardous waste.

Like crystalline silicon, the main environmental risks involved in thin-film silicon production (a-Si) revolve around the chemicals used in their manufacture, in particular the toxic and highly flammable silane gas (SiH_4). Silane has been a component in a number of industrial accidents in semiconductor and photovoltaic plants, at least one of them fatal.

Similarly the principal risks of the other thin-film technologies lie in the toxic or flammable gases used in their manufacture, or in their safe disposal. These can be tackled through a mixture of good industrial practice, and appropriate take-back and recycling schemes.[3]

Cadmium telluride cells (CdTe)

Perhaps the solar technology that has come under most scrutiny and which raises the most health and environmental concerns is cadmium telluride thin-film. Cadmium is a highly toxic metal, with carcinogenic properties. Currently the majority of cadmium is used in paints, pesticides, stabilisers and batteries, although its use in solar applications has soared in recent years. While the use of cadmium is restricted in many circumstances, its use in solar technology is deemed safe, since it is bound in a stable, non-soluble form with tellurium. Nonetheless regular monitoring of staff employed in manufacturing facilities is required to ensure that there is no build up of toxicity.

Over the years a number of concerns have been voiced about the potential for cadmium to leak from CdTe panels in the event of a fire. Tests, however, have shown that the temperatures achieved in normal house fires are not hot enough to reach the melting point of CdTe (1050 °C). Experiments using other fuels designed to give higher temperatures have shown that the cells can melt, causing CdTe to run to the edge of the unit, raising concerns about leakage in the event of an industrial fire. Further tests involving complete CdTe modules though have shown that the liquid CdTe is captured in the molten glass of the module, rendering it essentially harmless. Thus while in theory there is the potential for some CdTe release, the majority would be captured in inert glass droplets. Although there has been a lot of attention on cadmium in solar applications, the industry is keen to point out that there is 2500 times more cadmium in a small nickel-cadmium battery than in a standard CdTe module, and these are far less resistant to fire.

As always it is important to put the dangers of cadmium in solar cells into some kind of context. Despite the real risks associated with the release of cadmium into the environment, it is worth remembering that under normal circumstances, coal power stations release about 360 times more cadmium into the environment per MWh than a CdTe module.[4]

Water consumption

Of the main environmental concerns that have been expressed, it is water use that is likely to cause the biggest conflicts in the near term. Water in deserts is scare, and power generation can use a lot of it. This is true of many forms of power generation, but uniquely, large-scale solar applications must be sited according to where the solar resource is. They cannot mine it and transport it to a less water-stressed area, as can fossil forms of electricity generation.

Water use in solar applications is divided into three parts: water for cooling, water for the generation process and water for washing the mirrors or panels. Of these, water for cooling is of most concern, since it involves the largest quantities.

The exact amount of water used depends largely on the technology that is being employed. PV, for example, requires very little water, just what is needed to wash the panels. Of the CSP technologies, linear Fresnel reflectors have the highest demand, followed by parabolic troughs, power towers and dish–Stirling systems. Table 7.1 shows the water demands of various solar technologies in comparison with various other forms of generation. It should be noted that like PV, dish–Stirling systems use only a little water for mirror washing.[5]

The reasons for the difference in cooling demands between LFR, parabolic trough and power tower systems have been outlined in the technology section, as have some of the alternative cooling technologies, but the basic principle is that the lower the input temperature into the generator, the cooler the output must be in order to maintain efficiency. Thus, power towers, which heat their transfer medium to a very high temperature, can tolerate a higher output temperature from the generator, reducing their cooling needs. If the balance between the input and output temperature is not maintained, the generator can lose efficiency.[6]

Of course there are alternatives to using water cooling, and again these have been outlined in the technology chapter. Broadly speaking dry-cooling systems can dramatically reduce water use, but they also increase costs (by 7–9 per cent in the case of installations in the Mojave Desert). These losses

Table 7.1 Water use of conventional and CSP technologies, including washing and cleaning

Technology	Cooling	Gallons / MWh	Performance penalty	Cost penalty
Coal / Nuclear	1. Once-through 2. Recirculating 3. Air cooling	23,000–27,000 400–750 50–65		
Natural gas	Recirculating	200		
Power tower	1. Recirculating 2. Hybrid 3. Air cooling	500–750 90–250 90	 1–3% 1.3%	 5%
Parabolic trough	1. Recirculating 2. Hybrid 3. Air cooling	800 100–450 78	 1–4% 4.5–5%	 8% 2–9%
Dish–Stirling engine	Mirror washing only	20		
Linear Fresnel reflector	Re-circulating	1000 (estimated)		

Source: *Concentrating Solar Power Commercial Application Study: Reducing Water Consumption of Concentrating Solar Power Electricity Generation*, Report to Congress, US Department of Energy. http://www1.eere.energy.gov/solar/pdfs/csp_water_study.pdf

may be lower in other areas, as the efficiency of dry-cooling systems depends greatly on the ambient air temperature. Cooler and higher deserts will have fewer problems than hot ones.[7]

A quick look at Table 7.1 shows that the water demands of the two most popular CSP technologies – power towers and parabolic troughs – are broadly equivalent to, or less than coal or nuclear power stations for an equivalent form of cooling. Only natural gas has a generally lower water consumption per MWh. Linear Fresnel reflectors have a higher water demand, although as this technology is much less advanced than the other CSP technologies it may be too soon to tell. Dish–Stirling systems require no cooling, only mirror washing and so have minimal water consumption. For any of these technologies the use of alternative cooling methods will dramatically reduce water consumption.

Building 10,000 MW of new parabolic troughs using standard recirculating cooling may place unbearable strain on a region's water resources if it is purely additional to all the other activities of the region. If, however, the new CSP plants are replacing a series of old coal or nuclear power stations that have reached the end of their life, then the net water consumption may be broadly equal.

Providing 15 per cent of Europe's energy demands from CSP in North Africa (see section on DESERTEC, Chapter 6) might require CSP plants producing 700 TWh of electricity. Using parabolic troughs fitted with hybrid cooling technology this would require some 70–315 billion gallons of water per year. While this sounds like a lot, at its highest it amounts to just 1.1 billion m^3 per year, equivalent to 0.33 per cent of the MENA region's current water consumption.[8] Given the potential that CSP could offer for water 'production' through desalination, and the low population density in much of the region, fears over water shortages hampering renewable energy development in the Middle East and North Africa are perhaps not as serious as some reports have claimed.

Since there is nothing uniquely water-hungry about solar technology for electricity generation, the issue centres on if there is enough water in the immediate areas where they must be sited, and what the options are. In the South-Western United States there have already been rumblings of discontent from some members of the agricultural community over the predicted boom in solar power, and it is easy to see their concerns. A recent report in the New York Times highlighted one situation in the Amargosa Valley in Southern Nevada. This is the site of a proposed 500 MW parabolic trough plant under development by Solar Millennium. Figures quoted in the paper give the expected water consumption as 1.3 billion gallons per year (assuming a water consumption of 800 gallons/MWh, this works out at a capacity of 37 per cent). This, however, is reported to constitute around 20 per cent of the region's water supply, and Solar Millennium is currently in discussions with local farmers to purchase water rights.[9] It is worth noting that using a dry-cooling system would reduce the water demand to just 25 million gallons per year, less than 2 per cent of the wet-cooling mechanism. Should buying water rights prove too difficult or expensive it seems likely that the plant will be forced to use a less water intensive cooling system, or locate elsewhere. In this way a market for water rights could help to alleviate some of the concerns of farmers. At the same time though it may act as a serious barrier to the development of CSP and the common benefits it could bring, meaning additional tools or support may be needed.

Regulatory change may be required in some areas to free up water resources. At present in California it is illegal to use drinking-quality water for cooling purposes, and this places renewable energy generators in direct competition with local agriculture. Currently there is a bill before the state's congress that may amend this process to ease the development of these technologies.

While developed countries like the United States may have a decent regulatory infrastructure for safeguarding and issuing water rights, the same cannot be said of all other countries. In crowded regions like India, where millions of people lack property rights and are reliant on informal access to natural streams and rivers for water, developers of CSP technology will need to be very careful over how they use the resource. It does not take much imagination to see a situation that could bring renewable energy developers in direct confrontation with human rights activists, conservationists and antipoverty campaigners over water access, a situation that is likely to benefit no-one and which would greatly harm the image of important technologies (not to mention the people or forests losing their water supply!). The subcontinent is already littered with examples of industrial interests over-exploiting water resources, causing lakes and rivers to run dry. Whether these companies are legally entitled to the water is not really the question. It seems likely that in many parts of the world simply being given the legal right to access water resources by local authorities will not be enough to constitute 'due diligence' from renewable energy developers, and they would be wise to seek local advice before pursuing potentially controversial projects.

The bottom line is that yes, certain types of solar electricity generation do consume significant quantities of water, and this must be minimised and considered. At the same time, however, there is nothing unique about solar in this regard and the water requirements are minimal compared with those of other users. There is no scope in this book to give a more comprehensive assessment of the water issue, other than to say that it will depend on the specific region where the solar is under development. If a region is so water stressed that there is simply no water available, then solar projects will need to be sited elsewhere, or restricted to those technologies with minimal demands.

Land use

There is no denying that large-scale solar installations will require lots of land. Figures for supplying the whole United States come up with anything from 12,000 to 40,000 km^2 depending on which studies you listen to. The DESERTEC Initiative aims to supply North Africa, the Middle East and a significant percentage of Europe's needs, and again this will require more than 15,000 km^2 [10] (of which it is estimated 6100 km^2 will be for Europe, including 3600 km^2 for transmission lines).[11, 12] This is big stuff. As someone who has grown up in one of the most overcrowded regions of the world, where the mere idea that there are thousands of square kilometres of empty land available for industrial exploitation seems laughable, it can be hard to be objective about the availability of land for solar. Indeed, alongside renewable energy, one of my major passions is wildlife and the expansion and 're-wilding' of open spaces. Set against this the thought of anything taking up yet more precious space makes my blood run cold. I fully understand the concerns of people who live in the desert and are dismayed to suddenly find that the lands around them are in demand. Yet a quick look at the figures shows that a sense of perspective must be maintained. Deserts tend to be big. They also tend to be empty. Furthermore the amounts of land required are not as unprecedented as they might seem.

Taking the world's second largest polluter – the United States – as the first example, what amount of land would actually be needed to supply a significant amount of its electricity from solar? Obviously to some extent this depends on the technology, but estimates from CSP developers have ranged from 23,000 to 28,000 km², [13] depending on the system in question. Taking the median point, and assuming that 25,000 km² of desert could supply the electricity consumption of the entire USA, this would involve the exploitation of around 6.6 per cent of the Mojave and Sonoran deserts, or just 2.8 per cent of total desert land available in the USA. For the country to supply 50 per cent of its needs would then require just 1.4 per cent of desert lands. It is probably fair to say that it might take somewhat more, since the installations would not be localised in one place, but spread out, with their own power lines, substations and access roads, but equally it may be the case that improvements in efficiency could counter-balance or even reduce the need for land. Furthermore, there is no particular reason why these facilities will need to be built in the remote 'wilderness' areas, many will be located near to existing towns and cities, forming industrial solar satellites.

In North Africa, the 6100 km², including transmission, required to supply 15 per cent of Europe's electricity planned by DESERTEC and outlined in the DLR reports, constitutes just 0.6 per cent of the land mass of Egypt, and set against the scale of the Sahara as a whole, with its 8.6 million km², it becomes a truly tiny fraction (0.07 per cent). If 15,000 km² were used to supply the majority of the MENA region with electricity too, this would still only rise to 0.17 per cent of the Sahara, never mind the whole region.

It is also useful to compare these land-use requirements with those of coal power generation. Estimates suggest that across the USA, coal mining has disturbed more than 23,000 km² and currently affects a minimum of 4200 km².[14] Factor in the trains, power stations and storage yards and the number will be even higher. Given that coal accounts for around 48 per cent of the US energy mix and the difference in land-use demands between coal power and CSP is perhaps not as great as it may seem (although to be fair, about 5 per cent of this coal is exported).[15] Indeed, according to the German Aerospace Centre (DLR), the land-use requirements of solar energy are actually lower than those of coal or nuclear power, once generation, mining, transportation and transmission are included. Furthermore the land exploited for coal power generation is much more heavily affected that that for solar power. It is dug up, blasted, poisoned and degraded. Solar is merely an unsightly addition to the landscape, something that is bolted on the surface, and which can be easily removed if it is no longer necessary. Hydropower is another interesting example. The 2000 MW Hoover Dam necessitated the creation of a 650 km² reservoir.[16] That amount of land in the US Southwest devoted to solar would likely produce many times more electricity than the Hoover Dam.

The coal example is illustrative in other countries too. As has been mentioned elsewhere, in the State of Jharkhand in India, some reports suggest that coal mining affects some 4000 km².[17] Looking down at the region on

Google Earth reveals a scarred and blackened landscape, and even the rivers are grey. By comparison, 4000 km² of CSP or PV in the Thar Desert could supply around 60–70 per cent of India's current electricity demand, with almost none of the environmental consequences associated with coal.

In fact according to the DLR, CSP has land-use requirements of around 6–10 km² / TWh (6–8 km² / TWh for parabolic troughs). This compares favourably with the other estimates made in the DESERTEC reports, which gives average land-use figures of 46 km²/ TWh for wind power, 50–100 km² / TWh for oil and coal steam cycle generation.[18] The same study suggests Europe could supply 65 per cent of its electricity from renewable energy using 1 per cent of its landmass. If this were accompanied by an equivalent reduction in coal and nuclear generation and associated infrastructure, there might be little net difference in total land requirements for energy use.[19]

When discussing land availability, it is also interesting to look at some of the other uses to which land is currently put. This is important, because in general people (myself included) seem to be poor at gauging scale, and in particular the scale of human activity. Tell people we need to install one hundred million photovoltaic panels to make a big impact on a region's energy mix and they will be shocked at the size of the enterprise, yet in one year alone – 2007 – more than 270 million personal computers were sold worldwide.[20] The same is true of land use. Again using the USA as an example, some current land uses include: 127,000 km² on lawns, 40,000 km² on irrigated corn, 51,000 km² on Department of Defence installations and 411,000 km² on protected areas for wildlife (national parks, refuges and wilderness areas).[21]

None of this is to say that land use must not be minimised and conducted as sensitively as possible. Whatever the comparisons may be, land is still a precious resource. I suspect that a lot of the fears about land use come from the sense that this is yet more land that is being drawn into intensive human use – something that is all too often a one-way street. Once land is used for one industrial purpose, it is often considered fair game for other purposes. Perhaps, in the future, we need to start looking more seriously at how we can give up land that is no longer needed, and not just assume that it can automatically go from one human use to another as we keep using up more land indefinitely.

Resource constraints

As the solar industry has grown, numerous concerns have been raised about the availability of certain key resources – rare metals such as indium and gallium, or the ability of silicon manufacturers to keep up with demand. Some of these are genuine concerns, others less so. In any case it is welcome that the issue is under scrutiny as there is little point in placing undue hope in a technology that simply cannot provide for the future. We have already done that more than once, and it would unwise to continue doing so!

Indium and gallium

In the last couple of years there has been increasing concern about the supply of indium and gallium, and the impacts that any shortages might have on the future growth and viability of certain forms of PV. Indium in particular has attracted attention because it is not only a crucial ingredient in CIS and CIGS cells, but is also used as a coating and connector layer in many other kinds of thin-film solar cells. As recently as 2010 some writers have declared that these shortages will represent the death of the thin-film industry in the coming years, with one writer concluding that indium supplies could be used within 13 years.[22] Are such predictions realistic or alarmist? Perhaps unsurprisingly the main suppliers of indium and gallium think that resource constraints will not be a problem for many years.

To begin with, they explain that solar panels utilise only a minority of the indium produced worldwide each year. Currently nearly 80 per cent of the global indium demand is consumed by flat panel displays (flat-screen televisions, computer monitors and the like).

According to industry analysts indium occurs at concentrations of around 0.05 ppm in the continental crust – higher than that of silver. In 2009, total production of indium amounted to around 500 tonnes (up from around 70 tonnes in 1989), with the metal now being recovered as a by-product from a wide range of base metals (such as zinc). A study conducted by the Indium Corporation (which controls much of the global indium trade) estimated that the proven global reserves of indium amount to about 50,000 tonnes, or 100 years of production at current rates. Of this, around 26,000 tonnes is located in the 'west' while another 23,000 tonnes is located in China and the former Soviet Union. This may turn out to be an important factor as competition for resources between the major industrial powers looks likely to increase in the coming decades (witness China's current restrictions on the export of rare earth metals).[23]

It is important to realise that because indium is mined alongside other metals, and then refined to produce pure indium, the availability of indium metal has as much to do with the means of production as the overall availability of the element. Currently, of the estimated 1500 tonnes of indium extracted from the ground worldwide, only around a third is actually refined (500 tonnes). This is an increase on previous years when this recovery yield was generally around 20 per cent. As demand for indium increases, it seems likely that it will become economical to recover even larger percentages of the metal. At the same time industrial processes to recover indium are improving, with waste indium from flat panel or solar cell manufacturing being recovered at an increasing rate and reducing the overall demand for 'virgin' indium.[24]

As with indium, gallium is produced as a by-product of other mining activities, and is extracted from aluminium ore (bauxite) as part of the aluminium refining process. The primary uses of gallium are in gallium-arsenide wafers for circuit boards in mobile phones and other portable communications devices. LED displays are the second largest consumer, followed by solar cells, batteries and alloys.

Currently only about 10 per cent of the gallium associated with bauxite is actually purified and recovered, given tremendous potential for expansion in production. Furthermore the element itself is found throughout the earth's crust, meaning there are unlikely to be any absolute supply shortages in the foreseeable future. Recent fluctuations in the price of gallium have been caused by short-term supply and demand issues, rather than by an intrinsic scarcity. The record high price recorded in 2001 is reported to have been due to stockpiling by mobile phone manufacturers. When a shortage did not materialise this led to a price crash. A second round of price rises in 2007 was again due to fears over short-term supplies while new refining facilities were developed.[25]

It is these sorts of price rises that will represent the challenge for PV industries reliant on rare elements. Even if there is no intrinsic shortage of the metal, resource bottlenecks and competition for the available stocks could lead to price rises and volatility, making these technologies less competitive with other forms of solar or renewable energy. As is the case with all industries, however, only time will tell which designs and technologies become the most successful.

So is there a need to be concerned about the availability of indium and gallium? Naturally the Indium Corporation thinks not – but are they correct? Certainly there is unlikely to be any absolute shortage in the next few years, although as we have seen, fluctuations in price could have an impact. Looking to the longer term though it still seems possible that there could be problems. If solar currently consumes 100 tonnes of indium per year, and there are 50,000 tonnes of reserves, we should have 500 years at current usage. But indium-based solar panels are still the minority, accounting for less than 1 GW of production per year. Throughout this book we have discussed the need for hundreds, if not thousands of GW of solar power in the next 50 years. At those sorts of production levels it seems likely that a shortage of indium would ensue, and it may not be economic to extract it at these rates. Fortunately there are many new and emerging photovoltaic technologies waiting in the wings, and we must hope that these are ready to be deployed before the existing technologies run into difficulty. Plus, there is always silicon.

Silicon

As an example of an effective resource constraint, in recent years silicon shortages have been a strong limiting factor on the growth of the photovoltaic market. Crystalline silicon modules make up around 80 per cent of the PV market, and crystalline silicon makes up around 45 per cent of the cost of a silicon PV module. Thus any fluctuation in the price of silicon can have a major impact on the economics and cost-effectiveness of PV.

Since silicon is the second most abundant element on earth, and is easily obtainable from sand or quartz, there are no absolute shortages in the amount that is available. However, for solar applications (or semiconductors) exceptionally pure silicon is needed, and is generally produced by 'growing' silicon crystals. These may be made of one single crystal (monosilicon) or many crystals (polysilicon). Thus there is a limit to the amount of solar grade silicon that is available at any one time.

In its early years, solar was a relatively small-scale affair and could survive on the 'off-cuts' of the far larger semiconductor industry. In the last decade, however, it has grown quickly, by almost 41 per cent per year from 2003 to 2008. Despite this there was little investment in new production capacity, perhaps because the new industry still seemed so uncertain. By 2004, demand for silicon outstripped supply, and by 2005 demand was 30 per cent greater than supply. The net result was that prices for solar silicon rose from $24–30/kg in 2003[26, 27] to up to $450/kg on the spot market 2008.[28] Fortunately, since 2008 a large amount of additional silicon capacity has come on stream and the price has begun to decline rapidly, down by 70 per cent in 2009,[29] and now around $78/kg in early 2011.[30] While the solar industry continues to grow, so too does the silicon manufacture capacity (currently by around 80 per cent per year). The long-term impacts that this low price will have on the cost of PV are not yet clear, but already the installed costs of systems have fallen and look likely to continue doing so in the future.

So it seems that for now the 'shortage' of silicon appears to have passed, and assuming that sufficient investment in new silicon supply can be maintained alongside the growing solar market, there is no reason to believe that it will be a long-term problem. One of the interesting side effects of the silicon shortage in 2004–2008, however, was a surge in interest in thin-film technologies, as manufacturers sought ways of minimising their exposure to high silicon prices. The fact that solar technology is not reliant on just one type of technology bodes well for its ability to overcome future supply constraints.

Greenhouse gas balance, and atmospheric pollution

One of the most crucial questions for any form of renewable energy is 'how much energy does it produce in relation to how much is put in?' Anyone who reads the online comments sections on articles about renewable energy will know that there are numerous rumours that the energy inputs into making photovoltaics are so high that they take many years to pay them back, or perhaps never. If this were true then it would obviously seriously undermine the case for photovoltaic energy. Fortunately these rumours appear to be unfounded.

The main energy inputs into the production of photovoltaic panels include the mining and extraction of raw materials, the production of crystal silicon ingots (in the case of crystalline silicon cells) and the manufacture and shipment of modules. Of these activities, it is the manufacture of silicon crystals that is the most critical, accounting for 45 per cent of the energy inputs for polysilicon panels. While the energy payback times for photovoltaics differ depending on the different technology types, numerous studies have estimated the energy payback time for crystalline silicon and thin-film solar panels (CIS) at 1–3.5 years, depending on the exact process and technology. NREL for its part estimates the energy payback time at 1–4 years for a range of PV types.[31] Naturally this will vary depending on a range of factors, including many outside the control of the solar manufacturer, but given an average lifetime of 30 years, the panels can be expected to pay back the energy used to manufacture them many times over; studies for crystalline silicon put this at 9–17 times. Indeed, as production scales increase and new and more efficient manufacturing techniques are introduced, this payback time can be expected to decrease significantly. Emerging thin-film technologies are also expected to have an even more favourable payback period.

Importantly, as the proportion of renewable energy in the grid increases, the energy payback should become ever more favourable. If renewable energy is being used to create more renewable energy generators, in theory a virtuous cycle should be created (i.e. if the amount of carbon dioxide per unit of electricity released by the national grid falls by 20 per cent, then the amount of carbon emitted by any industrial process getting its energy from the grid will also fall).[32, 33]

Aside from the reduction in greenhouse gas emissions, the production and use of photovoltaic panels also reduces the quantities of other important atmospheric pollutants released (compared with standard electricity generation). Using the United States energy mix as the comparison, producing 1000 kWh of photovoltaic electricity prevents the emission of 3.6 kg of sulphur dioxide, 2.3 kg of nitrogen oxides and 635 kg of carbon dioxide, compared with 'business as usual'.[34]

For concentrating solar thermal applications, the energy payback time is estimated to be even less than for photovoltaics. According to one study, the payback for a large CSP plant could be as low as 5–6.7 months, which, given the plant's lifespan of around 40 years, would mean a life-cycle emission of 8.5–11.3 g/CO_2/ kWh, depending on the technology.[35]

Conclusion

As with any technology there are drawbacks to the large-scale use of solar technology – fields buried under metal and glass, water courses diverted for cooling and increased mining activities for rare earth metals. Yet although these obstacles are real and must be handled in the best way possible, they pale into insignificance compared with the environmental drawbacks of most other energy sources. From carbon emissions to acid rain and land use, solar comes up trumps. So while advocates of solar must be aware of its limitations, they need not be afraid to stand up and defend its record.

Chapter 8

Conclusion

Writing this short book about solar power has been an interesting and exciting process. While the problems facing the world with respect to the environment, climate change and economic development are daunting, perhaps overwhelming, it is also hard not to get absorbed by the infectious optimism of the solar sector. From photovoltaic cells to software developments, tracking systems to new reflective surfaces, the technology is advancing rapidly and on a wide front. Yet this very dynamism imbuess any attempt to review the sector with a number of core contradictions. Solar power is expensive, but getting cheaper. Solar power is currently inconsequential in global energy terms, yet as the futurist Ray Kurzweil points out, if solar can continue its current growth rate, doubling every two years, then within 20 years it could meet 100 per cent of global energy needs.[1] Of course, that is a very big if, and there is no guarantee that solar will be able to maintain this stellar trajectory or emulate the success of the home computing or telecoms industries. This leaves us with a series of unanswered questions. Will the economics of solar electricity continue to improve to the point where it really does become the cheapest energy source available, and will the potential conflicts over land and water use stall this process? Will solar power emerge as a dominant energy source in the twenty-first century, and will large-scale energy from the desert play an important role in this revolution? Can this transition be achieved fast enough to mitigate the worst impacts of climate change? Who will be the winners and losers in this radical new energy world, and what will be the long-term consequences – cultural, environmental and geopolitical – of the shifting patterns of resource abundance and availability?

It might sound like a soft answer, but at this stage my personal instinct will only go so far as to say that solar energy *could* lead to the revolution outlined above, but that it is not certain that it will. The technology certainly works, and in coming years I am confident that with sufficient investment new breakthroughs in generation, transmission and storage will make large-scale solar ever cheaper and more practical, and in fact this process has already begun. Looking at the predictions of consultants like McKinsey and DLR it is easy to be optimistic. All confidently predict falling solar costs, with DLR even claiming that CSP will be the cheapest source of electricity and water in large desert regions by 2050. At the rate it is advancing, PV in the sunbelt regions should become increasingly competitive with centralised fossil fuel power generation long before that. The cost of thin-film solar has already dropped below $1 per watt according to some manufacturers.

While I love the optimism of these sorts of predictions, I also think a little more caution is necessary. Recent history is littered with ideas that seemed like good ones at small scale, but fell apart as they were extrapolated to the scale of human societies. Perhaps there are pitfalls in such large-scale solar that have not yet been considered? Will it be a shortage of rare earth metals, conflict over land or some other such pressure that we have not yet imagined?

Renewable energy and solar also have plenty of enemies and detractors, and there are groups who may actively seek to undermine the growth of renewable energy for their own purposes. Fortunately these have become fewer as time has gone on, but there are still a great many people with a vested interest in seeing these technologies fail. Decentralised energy in particular would appear to be a threat to the centralised model of control that power companies are used to, and while some will adapt to new conditions and develop new business models, others may seek to fight back and maintain the monopolies (and proft margins) they enjoy with their current control of the power supply.

Public and political apathy, confusion and misunderstanding could be powerful barriers as well, particularly in these early years when solar technologies still need public investment and support. Short-term political expediency could see governments shelving research programmes, cancelling projects or simply failing to take notice of a new emerging technology, preferring to throw their lot in with more established and easily understandable energy concepts. One example from my part of the world is the British government's infamous decision to cancel its indigenous, and in many ways cutting edge, space rocket programme of the 1960s and 1970s on the grounds that they saw no real future for satellite technology. Now, as then, we must hope that where one government stumbles, another will show more vision. Columbus was turned down by both the Genoese and the Portuguese before the Spanish finally funded his mission. There is a real hope too that new powers such as China, India and Brazil will invest in these technologies, developing them as an alternative to the fossil fuel driven economies they are seeking to supplant. At the same time, however, politicians seeking simplicity and powerful vested interests could channel investment into other plentiful, but dirty, energy sources such as coal with all the negative consequences that could entail.

My hope is that those reading this book will now understand a little more about the state of solar power today and the potential of the world's deserts to provide electricity for the world. There is a lot happening in this sector and I suspect that many people will be surprised to learn of the plans and plants that are being developed as we speak. I hope too that those who read this book will not judge it too harshly should it seem ludicrously out of date. This is a rapidly expanding area. So rapid in fact that it is almost impossible to keep up. Indeed if some of the predictions mentioned above are to come true this book almost has to be out of date by the time it is available to buy, and I sincerely hope it is. I would make one final plea to the renewable energy industry though. Just because this is a low carbon business does not mean we should ignore the other environmental consequences of our actions. If we get this right, it is a chance to begin reordering the global economy onto a fundamentally more sustainable and environmentally friendly path, and that is something surely worth striving for.

Notes

1   Introduction

1. European Photovoltaic Industry Association (EPIA), *Global Market Outlook for Photovoltaics Until 2014*, May 2010 update.
2. B. Worm *et al.*, 'Impacts of biodiversity loss on ocean ecosystem services', *Science*, 3 November, 2006: Vol. 314 pp. 787–790 (http://www.sciencemag.org/content/314/5800/787.abstract).
3. Dr Niles Eldredge, 2005, writing in 'Action Bioscience' Website, published by American Institute of Biological Sciences, http://www.actionbioscience.org/newfrontiers/eldredge2.html.
4. Danae S.M. Maniatis, *Ecosystem Services of the Congo Basin Forests*, 2008, Global Canopy Project.
5. World Health Organization (WHO) and UNICEF, *Water for Life – Making it Happen*, 2004.
6. Government of India, National Action Plan on Climate Change, pp. 32, 30 June 2008.
7. *The Guardian*, 'Volcano causes travel disruption around the world', 18 April 2010.
8. IPCC Fourth Framework Assessment, *Climate Change 2007*.
9. ibid.
10. Jeffrey D. Sachs, 'The end of poverty', *Time Magazine*, 14 March 2005, pp. 42–52.
11. Reuters, 'UK study shows 94 percent fish stock fall since 1889', 4 May 2010.
12. *USA Today*, editorial cartoon, 12 July 2009.
13. Tyndall Centre, http://www.tyndall.ac.uk
14. A report of Working Group 1 of the Intergovernmental Panel on Climate Change, *Summary for Policymakers*. http://www.ipcc.ch/pdf/assessment-report/ar4/wg1/ar4-wg1-spm.pdf.
15. Working Group 1 of the Intergovernmental Panel on Climate Change, *Summary for Policymakers*. http://www.ipcc.ch/publications_and_data/ar4/wg1/en/spmsspm-projections-of.html.
16. Working Group 3 of the Intergovernmental Panel on Climate Change, 'Mitigation in the long term (after 2030)'. http://www.ipcc.ch/publications_and_data/ar4/wg3/en/spmsspm-d.html.
17. International Energy Agency, *World Energy Outlook 2009*.
18. Earth System Research Laboratory of the National Ocean Atmospheric Administration, Spring 2010. http://www.esrl.noaa.gov/gmd/ccgg/trends/.

19 Pew Centre on Climate Change, *Electricity Overview*, 2009. http://www.pewclimate.org/docUploads/ElectricityOverview.pdf.
20 International Energy Agency, *World Energy Outlook 2004*. http://www.iea.org/textbase/nppdf/free/2004/weo2004.pdf.
21 US Energy Information Administration, 'Short term energy outlook'. http://www.eia.doe.gov/emeu/steo/pub/contents.html.
22 *Business Green*, 'IEA accused of deliberately undermining global renewable industry', January 2009. http://www.businessgreen.com/bg/news/1806340/iea-accused-deliberately-undermining-global-renewables-industry.
23 BBC News, 'BP faces down shareholder protest on oil sands project', 15 April 2010. http://news.bbc.co.uk/1/hi/business/8621366.stm.
24 R. Sandrea, 'OPEC's next challenge – rethinking their quota system'. *Oil and Gas Journal* Vol. 101.29 2003. http://www.ipc66.com/publications/OPEC_Nex_Challenge_quota_system.pdf.
25 International Energy Agency, *World Energy Outlook 2008*.
26 G. Monbiot, 'When will the oil run out', *The Guardian*, 15 December 2008. http://www.guardian.co.uk/business/2008/dec/15/oil-peak-energy-iea.
27 *Grist*, 'Natural gas from fracking is worse than coal says new study', 11 April 2011. http://www.grist.org/list/2011-04-11-natural-gas-from-fracking-is-worse-for-climate-than-coal-says-ne.
28 J. Schindler and W. Zittel, 'Alternative world energy outlook 2006: a possible path towards a sustainable future' in *Advances in Solar Energy, Vol. 17*, 2007, Earthscan Publishing.
29 ibid.
30 ibid.
31 NDA, Nuclear Decommissioning Authority Annual Report and Accounts 2007/08. http://www.nda.gov.uk/documents/upload/Annual-Report-and-Accounts-2007-2008.pdf.
32 BBC News, 'Nuclear clean up to cost £70bn', 30 March 2006. http://news.bbc.co.uk/2/hi/business/4859980.stm.
33 J. Schindler and W. Zittel, 'Alternative world energy outlook 2006: a possible path towards a sustainable future' in *Advances in Solar Energy, Vol. 17*, 2007, Earthscan Publishing.
34 *BP Statistical Review of World Energy*, June 2010.
35 International Energy Agency, *Key World Energy Statistics 2009*.
36 World Health Organization, Media Centre. http://www.who.int/mediacentre/factsheets/fs292/en/index.html.
37 World Health Organization, *The Energy Access Situation in Developing Countries*, 2009. http://www.who.int/indoorair/publications/Energy_Access_Report_Brief.pdf.
38 European Wind Energy Association.
39 Global Wind Energy Council 2009.
40 National Energy Technology Laboratory presentation 'Tracking new coal fired power plants', 12 July 2011, p. 17, using data from Platts – UDI WEPPDB. http://www.netl.doe.gov/coal/refshelf/ncp.pdf.
41 http://www.allbusiness.com/africa/892509-1.html.
42 K. Komoto, M. Ito, P. van der Vleuten, D. Faiman and K. Kurokawa, *Energy from the Desert: Practical Proposals for Very Large Scale Photovoltaics*, 2007, Earthscan.

2 Technology

1. US Energy Information Administration (EIA).
2. http://www.daviddarling.info/encyclopedia/A/Archimedes_and_the_burning_mirrors.html.
3. J. L. Hunt, 'A burning question'. http://www.mlahanas.de/Greeks/Mirrors.htm.
4. Renewable Energy World.
5. National Renewable Energy Laboratory. http://www.nrel.gov/csp/solarpaces/project_detail.cfm/projectID=3.
6. National Renewable Energy Laboratory. http://www.nrel.gov/csp/solarpaces/project_detail.cfm/projectID=38.
7. National Renewable Energy Laboratory, Trough Net.
8. US Department of Energy, *Concentrating Solar Power Commercial Application Study: Reducing Water Consumption of Concentrating Solar Power Electricity Generation*, Report to Congress.
9. ibid.
10. ibid.
11. ibid.
12. ibid.
13. National Renewable Energy Laboratory. http://www.nrel.gov/csp/troughnet/solar_field.htm.
14. ibid.
15. Eckhard Lüpfert, Michael Geyer *et al.*, 'EuroTrough: Design issues and prototype testing at PSA', presentation by at ASME Solar Energy: The Power to Choose, Washington, DC, April 21–25, 2001.
16. National Renewable Energy Laboratory. http://www.nrel.gov/csp/troughnet/solar_field.htm.
17. ibid.
18. ibid.
19. A Global Overview of Renewable Energy Sources (AGORES).
20. A Global Overview of Renewable Energy Sources (AGORES) 'INDITEP'. http://www.agores.org/General/INDITEP.htm.
21. Pers. comm., David Faiman
22. National Renewable Energy Laboratory. http://www.nrel.gov/csp/troughnet/solar_field.html.
23. *CSP Today*, 'Parabolic trough reflectors: Does glass still have the cutting edge?', 14 January 2010.
24. E-Solar website. http://www.esolar.com.
25. Solarpaces. http://www.solarpaces.org/CSP_Technology/docs/solar_tower.pdf.
26. National Renewable Energy Laboratory website, background information.
27. Stirling Energy Systems Press Release. http://www.stirlingenergy.com/press-room.htm.
28. Interview with Anil Srivastava, June 2011.
29. Slaich Bergmann website. http://www.sbp.de/en.
30. *The Namibian*, 'Green tower not all hot air', 20 February 2007.
31. European Photovoltaic Industry Association, *Global Market Outlook for Photovoltaic until 2015*.
32. *Renewable Energy World*, 'Spain generated 3 per cent of its electricity from solar in 2010', 28 January 2011. http://www.renewableenergyworld.com/rea/news/article/2011/01/spain-generated-3-of-its-electricity-from-solar-in-2010.
33. National Renewable Energy Laboratory, *Best Research-Cell Efficiencies*.
34. SolarPlaza.com, 'Top 10 world's most efficient solar PV modules: monocrystalline silicon', 8 July 2011.

35 SolarPlaza.com, 'Top 10 world's most efficient solar PV modules – polycrystalline silicon', 8 December 2011.
36 *Energy and Capital*, 'Solar cell and microchip costs set to decrease', 10 November 2011.
37 Kosuke Kurokawa (ed.), *Energy from the Desert*, Chapter 3, p. 37, 2003, Earthscan.
38 ibid.
39 National Renewable Energy Laboratory, *Best Research-Cell Efficiencies*.
40 Kosuke Kurokawa (ed.), *Energy from the Desert*, Chapter 3, p. 37, 2003, Earthscan.
41 A. Cameron, 'Staying safe: how the PV industry is minimizing the hazards of solar cell production', *Renewable Energy World*, November 2005.
42 National Renewable Energy Laboratory. http://www.nrel.gov/pv/thin_film/pn_techbased_copper_indium_diselenide.html.
43 SolarPlaza.com, 'Top 10 world's most efficient CI(G)S modules', November 2011.
44 *Renewable Energy World*, 'The rise of CIGS – finally?', 22 July 2010. http://www.renewableenergyworld.com/rea/news/article/2010/07/the-rise-of-cigs-finally.
45 National Renewable Energy Laboratory. *Best Research-Cell Efficiencies*.
46 Millennium Technology Prize media centre.
47 National Renewable Energy Laboratory. *Best Research-Cell Efficiencies*.
48 ibid.
49 ibid.
50 *Renewable Energy World*, 'Tracking the CPV global market: Ready to fulfil its potential?', 8 August 2011.
51 German Aerospace Centre (DLR), Institute of Technical Thermodynamics Section Systems Analysis and Technology Assessment, *AQUA-CSP: Concentrating Solar Power for Seawater Desalination*.
52 German Aerospace Centre (DLR), Institute of Technical Thermodynamics Section Systems Analysis and Technology Assessment, *MED-CSP: Concentrating Solar Power for the Mediterranean Region*.
53 German Aerospace Centre (DLR), Institute of Technical Thermodynamics Section Systems Analysis and Technology Assessment, *TRANS-CSP: Trans-Mediterranean Interconnection for Concentrating Solar Power*.
54 ibid.
55 ibid.
56 Physord.com, 'World first superconducting DC power transmission system a step closer', 8 March 2010. http://www.physorg.com/news187251385.html.
57 SustainableBusiness.com, 'Superconducting DC cables: Technology is ready', 3 March 2010.

3 Energy from the desert

1 Earth Summit+5, Special Session of the General Assembly to Review and Appraise Implementation of Agenda 21, *The United Nations Convention to Combat Desertification: A New Response to an Age-Old Problem*, 1997. http://www.un.org/ecosocdev/geninfo/sustdev/desert.htm.
2 Eden Foundation, 'Desertification – a threat to the Sahel', 1994. http://www.eden-foundation.org/project/desertif.html.
3 K. Komoto, M. Ito, P. van der Vleuten, D. Faiman and K. Kurokawa, *Energy from the Desert: Practical Proposals for Very Large Scale Photovoltaics*, 2007, Earthscan.

4 International Energy Agency, *Key Energy Statistics 2009*.
5 CSP-MED 2005, DLR and ECOSTAR 2005.
6 K. Kurokawa, (ed.) *Energy from the Desert: Feasibility of Very Large Scale Photovoltaic Power Generation (VLS-PV) System*, 2004, James and James.
7 International Energy Agency, *Key Energy Statistics 2009*.
8 CBS News, 'The most polluted places on earth', 8 January 2010.
9 *Der Spiegel*, 'China's poison for the planet', 2 January 2007. http://www.spiegel.de/international/spiegel/0,1518,461828-2,00.html.
10 *The Guardian*, 'Yangtze River dolphin driven to extinction', 8 August 2007. http://www.guardian.co.uk/environment/2007/aug/08/endangeredspecies.conservation.
11 US Department of Energy, *Energy Information Administration Statistics* (http://tonto.eia.doe.gov/cfapps/ipdbproject/iedindex3.cfm?tid=90&pid=44&aid=8&cid=regions&syid=2005&eyid=2009&unit=MTCDPP).
12 *Financial Times*, '750,000 a year killed by Chinese air pollution', 2 July 2007. http://www.ft.com/cms/s/0/8f40e248-28c7-11dc-af78-000b5df10621.html#axzz1D6hef9qq.
13 http://www.worldwatch.org/node/4985
14 R.K. Tiwary and B.B. Dhar, 'Environmental pollution from coal mining activities in Damodar River basin, India', *Mine, Water and the Environment*, Vol. 13 (1994), pp.1–10.
15 Website of the Dutch Embassy to Chile. http://chili.nlambassade.org/Producten_en_Diensten/Economie_en_Handel/Publicaties_en_Documenten_Afdeling_Handel_en_Economie/Nieuwsbrieven_Afdeling_Handel_en_Economie/Business_in_Brief_May_2010.html.
16 *Bloomberg Businessweek*, 'A solar mother lode for Chile's mines', 10 February 2011.

4 Economic and policy aspects of solar power, and the status of regional markets

1 Presentation by Ecofys, 'Solar urban planning: PV in urban environment, Germany', in *International Energy Agency Task Ten*. http://www.iea-pvps-task10.org/IMG/pdf/Experiencias-internacionales_Alemania_Ecofys_SLidner.pdf.
2 Solarbuzz. http://www.solarbuzz.com/going-solar/using/quick-facts.
3 Solar Power Top Tips – website. http://www.solarpowertoptips.com/solar-electricity.aspx.
4 *McKinsey Quarterly*, 'Evaluating the potential of solar technologies', June 2008. https://www.mckinseyquarterly.com/wrapper.aspx?ar=2426&story=true&url=http%3a%2f%2fwww.mckinseyquarterly.com%2fEvaluating_the_potential_of_solar_technologies_2426%3fpagenum%3d1%23interactive&pgn=evpo09_exhibit.
5 SolarBuzz Solar Energy Industry Price Index. http://www.solarbuzz.com/facts-and-figures/retail-price-environment/solar-electricity-prices.
6 *EE Times*, 'GE claims solar-cell breakthrough', 28 January 2008. http://www.eetimes.com/electronics-news/4076031/GSE-claims-solar-cell-breakthrough.
7 SolarBuzz Retail Price summary. http://www.solarbuzz.com/facts-and-figures/retail-price-environment/module-prices.
8 ibid.
9 *McKinsey Quarterly*, 'Evaluating the potential of solar technologies', June 2008. https://www.mckinseyquarterly.com/wrapper.aspx?ar=2426&story=true&url=http%3a%2f%2fwww.mckinseyquarterly.com%2fEvaluating_the_potential_

of_solar_technologies_2426%3fpagenum%3d1%23interactive&pgn=evpo09_exhibit.
10. SolarBuzz Retail Price summary. http://www.solarbuzz.com/facts-and-figures/retail-price-environment/module-prices.
11. *McKinsey Quarterly*, 'Evaluating the potential of solar technologies', June 2008. https://www.mckinseyquarterly.com/wrapper.aspx?ar=2426&story=true&url=http%3a%2f%2fwww.mckinseyquarterly.com%2fEvaluating_the_potential_of_solar_technologies_2426%3fpagenum%3d1%23interactive&pgn=evpo09_exhibit.
12. German Aerospace Centre (DLR), Institute of Technical Thermodynamics Section Systems Analysis and Technology Assessment, *AQUA-CSP: Concentrating Solar Power for Seawater Desalination*.
13. German Aerospace Centre (DLR), Institute of Technical Thermodynamics Section Systems Analysis and Technology Assessment, *TRANS-CSP: Trans-Mediterranean Interconnection for Concentrating Solar Power*.
14. *McKinsey Quarterly*, 'Evaluating the potential of solar technologies', June 2008. https://www.mckinseyquarterly.com/wrapper.aspx?ar=2426&story=true&url=http%3a%2f%2fwww.mckinseyquarterly.com%2fEvaluating_the_potential_of_solar_technologies_2426%3fpagenum%3d1%23interactive&pgn=evpo09_exhibit.
15. Reuters,'Court orders $507.5 million damages in Exxon Valdez spill', 15 June 2009. http://www.reuters.com/article/2009/06/15/us-exxon-award-idUSTRE55E6DU20090615.
16. *Federal Register*, 'Department of Commerce – Intent to Prepare and Environmental Impact Statement on the Exxon Valdez Oil Spill Trustee Council's Restoration Experts', Vol. 75, No. 14, Notices p. 3706, Friday, January 22, 2010. http://www.fakr.noaa.gov/notice/75fr3706.pdf.
17. International Centre for Technology Assessment, *The Real Price of Gasoline – Report No.3: An Analysis of the Hidden External Costs Consumers Pay to Fuel their Automobiles*, 1998. http://www.icta.org/doc/Real%20Price%20of%20Gasoline.pdf
18. ibid.
19. Speech by Achim Steiner, UN Under-Secretary General and Executive Director, UN Environment Programme (UNEP) at meeting of OECD Council at Ministerial Level on the Theme – *The Crisis and Beyond: building a stronger, cleaner, fairer world economy*, 24 June 2009. http://www.unep.org/Documents.Multilingual/Default.Print.asp?DocumentID=591&ArticleID=6233&l=en.
20. International Institute for Sustainable Development, *The Global Subsidies Initiative – Untold Billions: Fossil Fuel Subsidies, Their Impacts and the Path to Reform*, March 2010. http://www.globalsubsidies.org/files/assets/effects_ffs.pdf.
21. Reuters, 'G20 agrees on phase-out of fossil fuel subsidies', 25 September 2009. http://www.reuters.com/article/2009/09/26/us-g20-energy-idUSTRE58O18U20090926?pageNumber=1.
22. Ofgem, Information Note 4, February 2010. http://www.ofgem.gov.uk/Media/PressRel/Documents1/RO%20Buy-Out%20price%202010%2011%20FINAL%20FINAL.pdf.
23. Ofgem website. http://www.ofgem.gov.uk/Sustainability/Environment/RenewablObl/Pages/RenewablObl.aspx.
24. http://www.newenergyfocus.com/do/ecco/view_item?listid=1&listcatid=32&listitemid=3524
25. http://www.pointcarbon.com/, 8 February 2011.

26 CDM Watch Press Release, 'CDM Coal Power Projects violate Kyoto Protocol', 22 March 2010. http://www.cdm-watch.org/wordpress/wp-content/uploads/2010/03/pr_cdm-coal-power-projects-violate-kyoto-protocol_english.pdf.
27 *International Rivers*, 'Wikileaks cable highlights high-level CDM scam in India', 20 September 2011.
28 Environmental Investigation Agency, 'EU Moves to ban fake carbon credits', Press Release, 21 Jan 2011. http://www.eia-international.org/cgi/news/news.cgi?t=template&a=627&source=.
29 Environmental Investigation Agency, *HFC-23 Offsets in the Context of the Emissions Trading Scheme*, 2010. http://www.eia-international.org/files/reports199-1.pdf.
30 *The Guardian*, 'Coal price reaches new heights', 1 April 2011. http://www.guardian.co.uk/business/2011/apr/01/coal-price-reaches-new-heights.
31 *Global Feed in Tariffs,* 'Germany confirms reduced feed-in tariffs rates for 1st July', 27 April 2010. http://www.globalfeedintariffs.com/2010/04/27/germany-confirms-reduced-feed-in-tariff-rates-for-1st-july-2010/.
32 SolarBuzz Regional PV markets Europe. http://www.solarbuzz.com/facts-and-figures/fast-facts/regional-pv-markets-europe.
33 *Renewable Energy World*, 'Spanish PV after the crash', 29 April 2010. http://www.renewableenergyworld.com/rea/news/print/article/2010/04/spanish-pv-after-the-crash.
34 SolarBuzz Regional PV markets Europe. http://www.solarbuzz.com/facts-and-figures/fast-facts/regional-pv-markets-europe.
35 *Renewable Energy World*, 'Spanish PV after the crash', 29 April 2010. http://www.renewableenergyworld.com/rea/news/print/article/2010/04/spanish-pv-after-the-crash.
36 Solar Power and Chemical Energy Systems (SolarPaces), 'Legislation promoting CSP implementation'. http://www.solarpaces.org/Library/Legislation/legislation.htm.
37 ibid.
38 Ministry of Economic Development, Italian Ministerial Decree 11/04/2008: Criteria and methods to stimulate the production of electricity from solar source through cycles (Official journal of 30/04/2008 no. 101). http://www.solarpaces.org/News/docs/Decree%20Solar%20Thermal%20Power%20Italy%2011042008%20English.PDF.
39 SolarBuzz Regional PV markets Europe. http://www.solarbuzz.com/facts-and-figures/fast-facts/regional-pv-markets-europe.
40 Reuters, 'Italy to cut solar power incentives in 2011–2013', 9 July 2010. http://www.reuters.com/article/2010/07/09/italy-solar-idUSLDE66627420100709.
41 World Resources Institute, 'Bottom line of renewable energy tax credits'. http://www.wri.org/publication/bottom-line-series-renewable-energy-tax-credits.
42 *The Guardian*, 'Coal price reaches new heights', 1 April 2011. http://www.guardian.co.uk/business/2011/apr/01/coal-price-reaches-new-heights.
43 SolarBuzz Regional PV markets Europe http://www.solarbuzz.com/facts-and-figures/fast-facts/regional-pv-markets-united-states.
44 European Photovoltaic Industry Association, *Global Market for Photovoltaics until 2014* (May 2010 update).
45 *Eco Seed*, 'Huge Chinese PV production capacity seen in 2011', 19 November 2010. http://www.ecoseed.org/en/business/renewable-energy/article/95-renewable-energy/8457-huge-chinese-pv-production-capacity-seen-in-2011.
46 SEMI PV Group China, 'China's PV industry keeps growing', February 2010 http://www.pvgroup.org/NewsArchive/ctr_034481.

47 *Eco Seed*, 'Huge Chinese PV production capacity seen in 2011' 19 November 2010. http://www.ecoseed.org/en/business/renewable-energy/article/95-renewable-energy/8457-huge-chinese-pv-production-capacity-seen-in-2011.
48 SEMI PV Group China, 'China's PV industry keeps growing', February 2010 http://www.pvgroup.org/NewsArchive/ctr_034481.
49 *Green World Investor*, 'First solar's showpiece 2000 MW solar plant in Inner Mongolia gets stalled by Chinese protectionism', 16 August 2010. http://greenworldinvestor.com/2010/08/16/first-solars-showpiece-2000-mw-solar-plant-in-inner-mongolia-gets-stalled-chinese-protectionism/.
50 *Business Green*, 'China to set solar feed in tariff by year end, says Suntech chairman', 24 August 2009. http://www.businessgreen.com/bg/news/1806748/china-set-solar-feed-tariff-suntech-chairman.
51 SolarPACES, Greenpeace and ESTELA, *Concentrating Solar Power – Global Outlook 2009: Why Renewable Energy is Hot*.
52 *Renewable Energy World*, 'Renewable energy policy update for China', 21 July 2010. http://www.renewableenergyworld.com/rea/news/article/2010/07/renewable-energy-policy-update-for-china.
53 *Green World Investor*, 'First solar's showpiece 2000 MW solar plant in Inner Mongolia gets stalled by Chinese protectionism', 16 August 2010. http://greenworldinvestor.com/2010/08/16/first-solars-showpiece-2000-mw-solar-plant-in-inner-mongolia-gets-stalled-chinese-protectionism/.
54 First Solar, 'First Solar and China Guangdong Nuclear to develop Ordos Solar Project', press release 5 January 2011. http://investor.firstsolar.com/releasedetail.cfm?ReleaseID=573747.
55 SolarPACES, Greenpeace and ESTELA, *Concentrating Solar Power – Global Outlook 2009: Why Renewable Energy is Hot*.
56 *The Guardian*, 'Greenwash: The dream of the first eco-city was built on a fiction', 31 December 2008. http://www.guardian.co.uk/environment/2009/apr/23/greenwash-dongtan-ecocity.
57 Jawarhal Nehru National Solar Missions Towards Building Solar India.
58 Global Green USA, 'Solar Report Card India'. http://www.globalgreen.org/solarreportcard/India.pdf.
59 SolarBuzz Regional PV markets Asia-Pacific. http://www.solarbuzz.com/facts-and-figures/fast-facts/regional-pv-asia-pacific.
60 Jawaharlal Nehru National Solar Missions Towards Building Solar India.
61 CIA World Factbook 2011.
62 SolarPACES, Greenpeace and ESTELA, *Concentrating Solar Power – Global Outlook 2009: Why Renewable Energy is Hot*.
63 The Infrastructure Consortium for Africa, 'Solar energy being developed in Algeria', 2 December 2009. http://www.icafrica.org/en/news/infrastructure-news/article/solar-energy-being-developed-in-algeria-526/.
64 Magharebia.com, 'A new national energy strategy aims to create 100,000 green jobs in Algeria over 20 years', 8 February 2011.
65 Reuters, 'Morocco unveils $9 billion solar scheme', 3 November 2009. http://af.reuters.com/article/investingNews/idAFJOE5A202D20091103.
66 RenewableEnergyFocus.com, 'Isofoton installs PV in Morocco', 22 January 2009. http://www.renewableenergyfocus.com/view/1549/isofoton-installs-pv-in-morocco-/.
67 SolarPACES, Greenpeace and ESTELA, *Concentrating Solar Power – Global Outlook 2009: Why Renewable Energy is Hot*.
68 ibid.

69 *The Guardian*, 'Egypt plans 100 MW solar power plant', 12 July 2010. http://www.guardian.co.uk/environment/2010/jul/12/egypt-solar-power.
70 SolarPACES, Greenpeace and ESTELA, *Concentrating Solar Power – Global Outlook 2009: Why Renewable Energy is Hot*.
71 RenewableEnergyFocus.com, 'Middle East to build first major power plant' , 9 June 2010. http://www.renewableenergyfocus.com/view/10089/middle-east-to-build-first-major-solar-power-plant/.
72 SolarPACES, Greenpeace and ESTELA, *Concentrating Solar Power – Global Outlook 2009: Why Renewable Energy is Hot*.
73 *Renewable Energy World*, 'Israel to set up feed-in tariff, European bank to invest €100m in Israeli CSP plant', 18 February 2010. http://www.renewableenergyworld.com/rea/news/article/2010/02/israel-to-set-up-feed-in-tariff-european-bank-to-invest-100m-in-israeli-csp-plant.
74 SolarPACES, Greenpeace and ESTELA, *Concentrating Solar Power – Global Outlook 2009: Why Renewable Energy is Hot*.
75 *The Guardian*, 'South Africa solar power plant', 25 October 2010. http://www.guardian.co.uk/environment/2010/oct/25/south-africa-solar-power-plant.
76 Institute of Sustainable Futures, *Subsidies that Encourage Fossil Fuel Use in Australia*, Working paper CR2003/01, 2003. http://www.isf.uts.edu.au/publications/CR_2003_paper.pdf.
77 Areva.com.

5 Existing and planned projects

1 SolarBuzz Market Share. http://www.solarbuzz.com/facts-and-figures/markets-growth/market-share.
2 *Renewable Energy World*.
3 US Department of Energy, Concentrating Solar Power 2007 Funding Opportunity Project Prospectus. http://www1.eere.energy.gov/solar/pdfs/csp_prospectus_112807.pdf.
4 Tony Seba blog. http://tonyseba.com/blog/.
5 http://www.eia.doe.gov/cneaf/electricity/epm/table5_6_a.html
6 Florida Power and Light (FPL) Project Briefing.
7 *Renewable Energy World*, 'DoE guarantees $1.45 billion loan for 250MW Abengoa Solar Thermal Project', 22 December 2010. http://www.renewableenergyworld.com/rea/blog/post/2010/12/doe-guarantees-1-45-billion-loan-for-250mw-abengoa-solar-thermal-project.
8 E-Solar Company website. http://www.esolar.com/our_projects/.
9 E-Solar Company website. http://www.esolar.com.
10 www.ivanpahsolar.com
11 California Energy Commission. http://www.energy.ca.gov/sitingcases/ivanpah/index.html.
12 *Techpulse*, 'Is Ivanpah the world's most efficient solar plant', 21 June 2010. http://techpulse360.com/2010/06/21/is-ivanpah-the-world%E2%80%99s-most-efficient-solar-plant-2/.
13 California Energy Commission, Energy Almanac. http://energyalmanac.ca.gov/electricity/levelized_costs.html.
14 California Energy Commission. http://www.energy.ca.gov/sitingcases/calicoolar/index.html.
15 California Energy Commission. http://www.energy.ca.gov/sitingcases/solartwo/index.html.

16 *Renewable Energy World*, 'Inside AREVA's Linear Fresnel solar power plant', 19 November 2010. http://www.renewableenergyworld.com/rea/news/article/2010/11/inside-a-linear-fresnel-solar-power-plant.
17 http://www.meteonorm.com/media/maps_online/gh_map_spain_hr.pdf
18 California Energy Commission. http://www.energy.ca.gov/sitingcases/ivanpah/index.html.
19 *CSP Today*, 'Lower cost of production is actually a by-product of Andasol's energy storage', 6 October 2008. http://social.csptoday.com/news/lower-cost-production-actually-product-andasol-1s-energy-storage.
20 National Renewable Energy Laboratory (NREL). http://www.nrel.gov/csp/solarpaces/project_detail.cfm/projectID=4.
21 US Department of Energy, Energy Information Administration Statistics. http://www.eia.gov/cfapps/ipdbproject/IEDIndex3.cfm?tid=2&pid=2&aid=2.
22 SolarPACES, Greenpeace and ESTELA, *Concentrating Solar Power – Global Outlook 2009: Why Renewable Energy is Hot*.
23 ibid.
24 MoroccoBoard.com, 'World Bank to back Morocco concentrated solar power plant', 14 September 2010. http://www.moroccoboard.com/news/34-news-release/1194-world-bank-to-back-morocco-concentrated-solar-power-plant.
25 Science and Development Network, 'Morocco invests $3.2 billion in renewable energy', 30 October 2008. http://www.scidev.net/en/news/morocco-invests-us-3-2-billion-in-renewable-energy.html.
26 Note that this is a fast changing situation, and the exact numbers will change as new installations are built and new markets, such as Italy and Czech Republic come to the fore.
27 PV Resources Top 50 PV power stations. http://www.pvresources.com/en/top50pv.php.
28 PV Resources 2010 large scale solar report.
29 *NewRop Mag* 'Moura (Amareleja) Photovoltaic Power Station', 17 September 2008. http://www.newropeans-magazine.org/content/view/8513/89/.
30 PV Resources – 1 March 2011.
31 http://www.pvresources.com.
32 hyyp://www.pvresources.com.
31 *Renewable Energy World*, 'GCL-Poly Builds 20-MW Solar PV Project', 4 January 2010. http://www.renewableenergyworld.com/rea/news/article/2010/01/gcl-poly-builds-20-mw-solar-pv-project.

6 Long-term visions

1 Solarbuzz.
2 US Department of Energy, Energy Information Administration Statistics. http://tonto.eia.doe.gov/cfapps/ipdbproject/iedindex3.cfm?tid=90&pid=44&aid=8&cid=regions&syid=2005&eyid=2009&unit=MTCDPP.
3 Masdar website. www.masdar.ae.
4 *Environment News Service*, 'Bush previews Abu Dhabi's planned carbon neutral, car free city', 14 January 2008. http://www.ens-newswire.com/ens/jan2008/2008-01-14-01.asp.
5 Wikipedia. http://en.wikipedia.org/wiki/Masdar_City#cite_note-meed-6.
6 *The Economist*, 'Clean Technology – Masdar Plan', 4 December 2008. http://www.economist.com/node/12673433?story_id=12673433.
7 Foster & Partners, website entry on Masdar. http://www.fosterandpartners.com/Projects/1515/Default.aspx.

8 Arabian Business.com, 'Abu Dhabi's Masdar plans to build second 100 MW solar plant', 17 February 2011. http://www.arabianbusiness.com/abu-dhabi-s-masdar-plans-build-second-100-mw-solar-plant-381329.html.
9 *Khaleej Times Online,* 'Completion of Masdar City pushed back', 10 October 2010. http://khaleejtimes.com/DisplayArticle09.asp?xfile=data/theuae/2010/October/theuae_October256.xml&section=theuae.
10 *Greentech Media Solar (GTM Solar)* 'More applied turmoil: Masdar PV abruptly changes management', 7 May 2010. http://www.greentechmedia.com/articles/read/more-applied-turmoil-masdar-pv-abruptly-changes-management/.
11 *Green World Investor,* 'Solar thin film technology sees weaker hands going out of business', 15 June 2010. http://greenworldinvestor.com/2010/06/15/solar-thin-film-technology-sees-weaker-hands-going-out-of-business/.
12 Masdar, 'Masdar City master plan review provides project update', press release 10 October 2010. http://www.masdar.ae/en/MediaArticle/NewsDescription.aspx?News_ID=150&News_Type=PR&MenuID=55&CatID=64.
13 Personal correspondence with Dr Gerry Wolff, DESERTEC UK.
14 Euractiv.com, 'EU CO2 emissions fell by 11% in 2009', 2 April 2010. http://www.euractiv.com/en/climate-environment/eu-co2-emissions-fell-11-2009-news-403298.
15 German Aerospace Centre (DLR), Institute of Technical Thermodynamics Section Systems Analysis and Technology Assessment, *MED-CSP: Concentrating Solar Power for the Mediterranean Region.*
16 German Aerospace Centre (DLR), Institute of Technical Thermodynamics Section Systems Analysis and Technology Assessment, *MED-CSP: Concentrating Solar Power for the Mediterranean Region.*
17 German Aerospace Centre (DLR), Institute of Technical Thermodynamics Section Systems Analysis and Technology Assessment, *TRANS-CSP: Trans-Mediterranean Interconnection for Concentrating Solar Power.*
18 ibid.
19 ibid.
20 German Aerospace Centre (DLR), Institute of Technical Thermodynamics Section Systems Analysis and Technology Assessment, *AQUA-CSP: Concentrating Solar Power for Seawater Desalination.*
21 German Aerospace Centre (DLR), Institute of Technical Thermodynamics Section Systems Analysis and Technology Assessment, *TRANS-CSP: Trans-Mediterranean Interconnection for Concentrating Solar Power.*
22 ibid.
23 Communiqué issued by the French Ministry for Ecology, Energy, Sustainable Development and Marine Affairs, Responsible for Green Technology and Climate Negotiations, 5 July 2010 – *Transgreen project: developing renewable energy in Africa and Europe.* http://www.ambafrance-uk.org/Transgreen-agreement-signed-in.html.
24 DESERTEC website.
25 http://libyaonline.com/news/details.php?id=14642.
26 Personal conversations with DESERTEC representatives.
27 Pedants will no doubt point out that the sun may not quite be the source of all energy on earth (some geothermal energy comes from radioactive decay, and tidal energy involves the gravity of the earth and moon), but the point still stands.
28 Jawaharlal Nehru National Solar Mission Towards Building SOLAR INDIA.
29 ibid.
30 *The Guardian,* 'India sets out ambitious solar plans to be paid for by rich nations', 4 August 2009. http://www.guardian.co.uk/environment/2009/aug/04/india-solar-power.

31 *Inside Climate News*, 'India's "Solar Deal", world's "most ambitious", not a done deal', 5 August 2009. http://solveclimate.com/news/20090805/india%E2%80%99s-solar-plan-world%E2%80%99s-most-ambitious-not-done-deal.
32 *Bloomberg Businessweek*, 'India's Copenhagen envoys unyielding on costs', 16 December 2009. http://www.businessweek.com/globalbiz/content/dec2009/gb20091216_417541.htm.
33 Jawaharlal Nehru National Solar Mission Towards Building SOLAR INDIA.
34 Government of Rajasthan Energy Department, Rajasthan Solar Energy Policy, 2010. http://www.rrecl.com/Rajasthan%20Solar%20Energy%20Policy%20-2010.pdf.
35 Merinews.com, 'Solar energy promises to keep India shining', 26 May 2009. http://www.merinews.com/article/solar-energy-promises-to-keep-india-shining/15770597.shtml.
36 DESERTEC Foundation website: DESERTEC India. http://www.desertec-india.org.in/cspthar.html.
37 *Solar India Online* 'Rajasthan government signed MOU with Clinton Foundation for solar power', January 2010. http://www.solarindiaonline.com/content/2010/01/rajasthan-govt-signed-mou-with-clinton-foundation-for-solar-power/.

7 Environmental and resource issues facing solar technology

1 RenewableEnergyFocus.com, 'Solar PV installation reached 17.5 GW in 2010', 18 January 2011. http://www.renewableenergyfocus.com/view/15219/solar-pv-installations-reached-175-gw-in-2010/.
2 *The Guardian*, 'Greenwash – How a wind farm could emit more carbon than a coal power station', 13 August 2009. http://www.guardian.co.uk/environment/2009/aug/13/wind-farm-peat-bog.
3 A. Cameron, 'Staying safe: How the PV industry is minimizing the hazards of solar cell production', *Renewable Energy World*, November 2005.
4 ibid.
5 US Department of Energy, *Concentrating Solar Power Commercial Application Study: Reducing Water Consumption of Concentrating Solar Power Electricity Generation*, Report to Congress. http://www1.eere.energy.gov/solar/pdfs/csp_water_study.pdf.
6 ibid.
7 ibid.
8 German Aerospace Centre (DLR), Institute of Technical Thermodynamics Section Systems Analysis and Technology Assessment, *AQUA-CSP: Concentrating Solar Power for Seawater Desalination*.
9 *New York Times*, 'Alternative energy projects stumble on a need for water', 30 September 2009. http://www.nytimes.com/2009/09/30/business/energy-environment/30water.html.
10 For this I am assuming around 9000 km^2 of CSP, generating 2500 TWh, based on 2500km^2 generating 700 TWh for Europe.
11 German Aerospace Centre (DLR), Institute of Technical Thermodynamics Section Systems Analysis and Technology Assessment, *MED-CSP: Concentrating Solar Power for the Mediterranean Region*.
12 German Aerospace Centre (DLR), Institute of Technical Thermodynamics Section Systems Analysis and Technology Assessment, *TRANS-CSP: Trans-Mediterranean Interconnection for Concentrating Solar Power*.

13 D. Mills and R. Morgan, 'Solar thermal electricity as the primary replacement for coal and oil in US generation and transport' in Sourcewatch.org, *Concentrating Solar Power Land Use*. http://www.sourcewatch.org/index.php?title=Concentrating_solar_power_land_use#cite_note-2.
14 Environment America Research and Policy Council, *On the Rise: Solar Thermal Power and the Fight Against Global Warming*, 2008. http://www.environmentamerica.org/uploads/0f/jZ/0fjZtsJDnQCqGKdr9a7Hjg/On-The-Rise.pdf.
15 US Energy Information Administration. www.eia.doe.gov/cneaf/coal/page/special/overview.html.
16 Environment America Research and Policy Council, *On the Rise: Solar Thermal Power and the Fight Against Global Warming*, 2008. http://www.environmentamerica.org/uploads/0f/jZ/0fjZtsJDnQCqGKdr9a7Hjg/On-The-Rise.pdf.
17 R. K. Tiwary and B. B. Dhar, 'Environmental pollution from coal mining activities in Damodar River basin', *India Mine Water and the Environment*, Vol 13, June–December Issue, 1994, pp.1–10.
18 German Aerospace Centre (DLR), Institute of Technical Thermodynamics Section Systems Analysis and Technology Assessment, *MED-CSP: Concentrating Solar Power for the Mediterranean Region*, (Chapter on Land use).
19 ibid.
20 Super Site blog, 'Mac market share for CY 2007 and Q4 2007' (quoting Gartner research), 22 January 2008. http://www.winsupersite.com/blogs/tabid/3256/entryid/74436/mac-market-share-for-cy-2007-and-q4-2007.aspx.
21 http://www.sourcewatch.org/index.php?title=Concentrating_solar_power_land_use#cite_note-2
22 Oilprice.com, 'Indium shortage creates problems for solar cell manufacturers', 8 October 2010. http://oilprice.com/Alternative-Energy/Solar-Energy/Indium-Shortage-Creates-problems-For-Solar-Cell-Manufacturers.html.
23 *PV Tech*, 'Do CIGS thin-film production expansions raise material supply concerns', 28 September 2009. http://www.pv-tech.org/editors_blog/do_cigs_thin-film_production_expansions_raise_material_supply_concerns.
24 ibid.
25 ibid.
26 China Nuvo Solar Energy Inc., industry information. http://www.chinanuvosolar.com/indinf.htm.
27 *Renewable Energy World*, '2005 Solar year-end review and 2006 solar industry forecast', 11 January 2006. http://www.renewableenergyworld.com/rea/news/print/article/2006/01/2005-solar-year-end-review-2006-solar-industry-forecast-41508.
28 *Renewable Energy World*, '10 Years in the sun: the most profitable decade in PV history draws to a close', 5 March 2010. http://www.renewableenergyworld.com/rea/news/article/2010/03/10-years-in-the-sun.
29 *Nature*, 'Plummeting silicon prices may boost solar sales', 5 August 2009. http://www.nature.com/news/2009/090805/full/460677c.html.
30 pvinsights.com.
31 National Renewable Energy Laboratory (NREL), *Energy Payback: Clean Energy from PV*. http://www.nrel.gov/docs/fy99osti/24619.pdf.
32 ibid.
33 K. E. Knapp and T. L. Jester, *An Empirical Perspective on the Energy Payback Time for Photovoltaic Modules*, 2000. Available at http://www.ecotopia.com/Apollo2/knapp/PVEPBTPaper.pdf.
34 National Renewable Energy Laboratory (NREL), *Energy Payback: Clean Energy from PV*. http://www.nrel.gov/docs/fy99osti/24619.pdf.

35 M. Z. Jacobson, 'Review of solutions to global warming, air pollution, and energy security', *Energy Environment Science,* 2009, **2**, 148–173.

8 Conclusion

1 *The 9 Billion,* '100% solar energy in 20 years no problem says Futurist Ray Kurzweil', 25 February 2011. http://www.the9billion.com/2011/02/25/100-solar-energy-in-20-yrs-no-problem-says-futurist-ray-kurzweil/.

Index

Information contained in a table is denoted by a bold page reference and information contained in a figure is denoted by an italicized page reference.

Abengoa Solar 101, **106**
Abu Dhabi *see* Masdar City, Abu Dhabi
AC (alternating current) 44, 56–8, 130
Acciona Energy of Spain 100, 114
Africa 4, 21, 60, **63**, 67; South Africa 13, **15**, 77, 93; Sub-Saharan 7, 67–8; *see also* DESERTEC; MENA (Middle East and North Africa)
agriculture 60, 144–5
air cooling systems 29–30, *29*, 53, 105, **149**
Algeria 28, 110, 132, 133–4; solar market 90; *see also* NEAL
alternating current *see* AC (alternating current)
Amargosa Solar Power Project **102**
Andasol projects 106, 107, **108**, *109*
Arab Spring 90
Arizona 85, 87, **102**; Maricopa 40, 98, **99**, 103
Arizona Public Service (APS) 101
a-Si (amorphous silicon) 47–8, 147
Asia **63**, 65–6, 79–80
Atacama Desert 68
Ausra 41, 94
Australia 43, 60, 93–5, **95**, 104; deserts 64–5; *see also* Ausra

back-up (natural gas) 21, 27–8, 97, 105; Israel 92, 112; Nevada 99
band-gap 48, 50
bauxite 154–5
Beacon Solar Energy Project **102**

biofuels 143–5
Blythe Solar Electric Power Plant **116**
Brightsource Energy 34, 93, 103, 105
business-as-usual scenario 128, **129**, 131, 137

cadmium 143, 147–8
cadmium telluride (CdTe) 47, 48–9, 53–4, 112; Cimarron PV *117*; health concerns 147–8; PV plants **116**
Calico Solar 41, **99**, **102**, 104
California 44, **63**, 101–4, **102**, **116**; cap and trade system 78, 86; electricity tariffs 72; parabolic trough power plants 27, 30–1, **32**, 36, *100*; power tower technology 37–9, 98, 101, 103, 105; Renewable Portfolio Standards 75, **85**, 87; Sunrise Powerlink 41; water resources 150
Canada 9, 49, 112, **113**, 118; pollution 14, **15**
cap and trade system *see* emissions trading
capacity, renewable energy production **132**
capacity factor (solar) 39, 61
capacity factors **24**, 39, 61
carbon dioxide emissions 14–16, **15**, 128–9, 157
carbon trading 74, 78–80
cars, electric 124–5
CBD (Convention on Biological Diversity) 6
CDM (Clean Development Mechanism) 78–80, 139; India 89; MENA (Middle East and North Africa) 90, 133; UAE 92
CdTe *see* cadmium telluride (CdTe)

cells 21–2, 44, 139, 147, 154–5; CdTe 49; CIGS cells 49–50, 154; CPV (concentrating photovoltaic) 52–3; disposal of 145–6; dye-sensitised 50–1; GaAs cells 51; thin-film technologies 47–8
central receivers *see* power towers
certificates, green 74–5
chemicals in cell manufacture 55, 146–7
Chile 68
China 3, 86–9, 113, 154, 160; indium 154; large scale PV 117–18; nuclear power 13, 14; pollution 14, 15, 65; PV solar installation rates 85; shale gas 11; wind power 17
China Guangdong Nuclear Solar Energy Development Co 88
CIGS cells 49–50
cities *see* Masdar City, Abu Dhabi
Clean Development Mechanism *see* CDM (Clean Development Mechanism)
climate change 4–6, 60, 80, 126–8, 159; India 89, 136
climate crisis 6–8
coal 49, 143, 152–3; Asia 65–6; Australia 93–4, 104; China 17; comparison with renewables 103, 129; economics of 74, 81, 133; future 129–30, 133, 160; India 148; relationship to water-use 149; South Africa 68
collectors 31, 33–6, 43
Combined Cycle Systems 25
competition 76, 81; in PV manufacturing 82; for resources 154–5
Concentrating Photovoltaics *see* CPV (Concentrating Photovoltaics)
Concentrating Solar Power *see* CSP (Concentrating Solar Power)
concentrators 51–2
consumers, cost of energy 10–11, 74, 76, 81
Convention on Biological Diversity (CBD) 6
conversion efficiencies: of CIGS cells 49; of CPV cells 52–3; of a-Si 47
cooling 28–30, 148, 149–50, **149**; dry-air cooling systems 53, 105, 149, 150; water resources 107, 131, 148

costs 46, 69–72, 71, 112, 115; CPV (concentrating photovoltaic) 52; feed-in tariffs 76, 82; of fossil fuels 72–4, 95, 117, 129, 133; future of renewables 129, 133, 156, 160; GaAs cells 51; of mirrors 37; of renewables 129; of transmission 57
CPV (Concentrating Photovoltaics) 52–4, 94
credits *see* emissions trading
crystalline PV 2, 46
crystalline silicon 45–6, 48, 112, 146, 156–7; large plants **116**
CSP (concentrating solar power) 18, 22–3, **24**, 69, 96–7; costs 70, 160; desalination 131, 150; feed-in tariffs 77, 88, 90, 93; Germany 82; land-use requirements 153; large scale installations 97–112; South Africa 93; Spain 104, 106–9; USA 97–104, 105; water use 29; *see also* DESERTEC; Dish-Stirling systems; Linear Fresnel reflectors
Czech Republic 77, 84, 85, 114–15

DC (direct current) 44, 56–7
desalination 54–5, 127, 131, 150
DESERTEC 120–1, 126–35, 152, 153
desertification 16, 144
deserts 59–62, **63**; availability 62–8
DeSoto Next Generation Solar Energy Center 115, **116**
developing countries 17, 78, 120, 128, 138
direct current *see* DC (direct current)
Direct Steam Systems (DISS) 35–6
disasters 9–10, 72–3
dishes *see* mirrors
Dish-Stirling systems 39–41
distribution 56–8
dry-cooling systems 29–30, 149–50
dye-sensitised cells 50–1

economic development 14–16
economics 10, 69–81, 159; business case for solar 96–7; CSP (concentrating solar power) 25, 37; photovoltaic cells (PV) 114, 156; thin-film technologies 51, 112
efficiency **24**, 30, 56, 148, 149; of cells 48, 51
Egypt 60, 92, 110–11

178 Index

electricity demand 45, 65–6, 152–3; global 14, 61
emissions 6–7, 14–16, 86, 128, 136; from standard electricity generation 157
emissions trading 76–80, 86
Emissions Trading Scheme *see* ETS (Emissions Trading Scheme)
energy efficiency *see* efficiency
energy transmission 56–8
environmental challenges of solar 142–53
eSolar 37, 101
ETS (Emissions Trading Scheme) 77–8, 80
Europe 16, 56, 70, 76, 81–4; large scale PV 113–15; renewable energy 17, 129–34, 153; *see also* ETS (Emissions Trading Scheme)
Eurotrough collector 33, 34
exports: of electricity 67–8; of solar cells 89
extinctions 4, 5
Exxon 72–3

feed-in tariffs 76, 77, 133; Algeria 90, 110; Australia 94, **95**; Europe 82–4, 98, 115; India 138; Israel 92–3, 111; South Africa 93; USA 84, 88
First Solar 48–9, 88, 115
Flabeg Solar International of Germany 36–7
Florida 87, 99, 100, 116
fossil fuels 4, 6, 16; economics of 81, 129, 130, 133; true cost 69, 72–4; *see also* coal; natural gas; peak oil
fracking 11
France **15**, 77, 84, 115

GaAs cells 51
gallium 49, 51, 53, 154–5
gas *see* natural gas
generators 3, 23, 29, 148
Germany 3, **15**, 72, 77, 85; large scale PV 112, **113**, 115; solar market 81–2
glass 36–7, 45, 146; *see also* mirrors
Gobi desert 65, 118
Granada 106, 107, **108**
Greece 77
green certificates 74–5
greenhouse gases **15**, 65, 156–7
grid, electricity 56–8, 62, 104
Gulf of Mexico 9–10, 72

Hawaii 7, 72, **87**, 99, 100
health concerns of solar 142–53
heat exchangers 21–2, 26, 29–30
heat storage system 38
heliostats 39, 101, 105
Hualapai Valley Solar Project Mohave 102
High Voltage Alternating Current (HVAC) 57
High Voltage Direct Current (HVDC) 57, 130
hybrid wet-dry cooling systems 30
hydrogen 55–6

IEA (International Energy Agency) 8–11, 81
Imperial Valley, California 40–1, **102**, 104
India 6–7, 65–6, 89, 150, 153; CDM project 79; coal 79, 153; feed-in tariffs 77; National Solar Mission 136–40; pollution 14, **15**; population 15, 54; PV installations 85
indium 49, 51, 153, 154–5
insolation 63
Integrated Solar and Combined Cycle system (ISCC) 25, 28, 99, 100, 110
Investment Tax Credit 84–5
Israel 77, 92–3, 112
Italy 72, 77, 83–4, 85; large scale PV 112, **113**, **114**; solar market 83–4
Ivanpah Solar Tower Project 105

Japan 3, **15**, 44–5, 77, 85; CIGS cells 50
Jawarhal Nehru, National Solar Mission 136–7, **138**
Jiangsu Operational **118**

Keahole Solar Power (Hawaii) 99
Kimberlina Solar Thermal Energy Plant (Arizona) 99
Kingman Solar Project (Arizona) 102
Korea, South 77
Kyoto Protocol *see* CDM (Clean Development Mechanism)

land-use requirements 151–3
lenses *see* mirrors

Index

Linear Fresnel Reflectors (LFRs) 23, 41–2, 104, 148–9
Luz system collectors 33–4

manufacturers 53, 139, 147; HFCs 80
markets 3, 45, **102**, 109, 112; *see also* feed-in tariffs
markets, regional 81–95
Masdar City, Abu Dhabi 118, 121–6, *123*
MED (multi-effect desalination) *see* desalination
Mediterranean 17, 44, 62; *see also* Plan Solar de Mediterranean
MENA (Middle East and North Africa) 66–7, 90–3, 150
methane 7, 11
Mexico 9, **15**, 54, 116
Middle East *see* MENA (Middle East and North Africa)
mirrors 21–5, 31, 36–7, 40–1, 52; Sun Tower 101–3; washing 148
modules (solar) 46, 52, 146–7; *see also* cells
Mojave Desert 37, 41, 103
Mongolia 65, 88
Morocco 28, **71**, 90–2, 110; *see also* MENA (Middle East and North Africa)
Munich RE 127

natural gas 11–12, 21, 27, 41, 133–4; Israel 92, 112; Ivanpah Solar Tower Project 103, 105; Nevada 99
NEAL (New Energy Algeria) 90
Negev Desert 93, 111
Nellis solar power plan **116**
Nevada 85, **87**, 99, **102**, 116
New Mexico 17, 85, **87**, 98, **116**, *117*
North Africa *see* MENA (Middle East and North Africa)
North America, deserts 62–4
nuclear power 12–14

oil 8–11, 72–3, 81, 129, 134

panels, solar 21, 36, 44, 54, 157; costs 70
parabolic troughs 30–6, 99–103, 107–9
Patagonian Desert 68
peak oil 8–12

photovoltaic cells (PV) *see* cells
photovoltaic technology (PV) 44–54, 85; large scale installations 112–19, 113, **116**, **118**
Plan Solar de Mediterranean (PSM) 120, 126–7, 135–6
Platforma Solara de Almeria, Spain 23–4
policy environment 72–4
pollution **15**, 78–9, 143, 148; Asia 65, 136; HFCs 80
Portugal 54, 77, 84, 112, 114
poverty 5, 15, 16
power towers 37–9, 98, 105, 109, **149**; India 137; Israel 112; planned **106**; Sierra Sun Tower 99, 101, 103; Spain 26, **108**
Production Tax Credit (PTC) 85
projects planned 96–119
PS10 and PS20 (Spain) 39, **106**, 107, 109, *110*
PSM *see* Plan Solar de Mediterranean
PV *see* photovoltaic technology (PV)

receiver *see* power towers
regional markets 81–95
regulatory changes, India 139
renewable energy 2–6, 16–17, *129*
Renewable Obligation Certificates (ROCs) 75
renewable obligations 74–5, 98
Renewable Portfolio Standard (RPS) 75, 85–6, 87
resource constraints 51, 153–6
reverse osmosis (RO) 55
Rice Solar Energy Project (California) **102**

Saguaro plant (New Mexico) 99–100
Sahara desert 66–7, 70, **71**; *see also* DESERTEC; MENA (Middle East and North Africa)
salt storage technology 26–7, 39
Saudi Arabia *see* MENA (Middle East and North Africa)
SEGS power plants (California) 32, 33, 36, 99, *100*
SES (Stirling Energy Systems) *see* Stirling Energy Systems
SES Solar One *see* Calico Solar
shale gas 11
Sierra Sun Tower (California) 101

silicon 44–7, 146–7, 155–6
solar cells *see* cells
solar chimneys 23, 42–3
solar energy, overview 18, 20–2
solar panels *see* panels, solar
Solar Tres (Spain) 39, **108**
solar updraft towers *see* solar chimneys
Solargenix collector *35*, 100
Solucar complex, Spain *110*
Sonoran desert 63, 101
South Africa 13, **15**, 77, 93; coal 68; solar market 93
South America, deserts 68
Spain 3, 82–3, 85, 91, 113–14; CPV (concentrating photovoltaic) 54; CSP plants 26, 104, 106–9, **106**, 108; feed-in tariffs 76, 77, 98, 133; Linear Fresnel Reflectors (LFRs) *42*; parabolic troughs 106–7; Platforma Solara de Almeria 23–4; power towers 38, *110*; PV installations 45, 46, *52*, 91, 112; solar chimneys *43*; wind power 16, 133
Stirling Energy Systems (USA) 40–1, **102**, 103–4
storage technology 26–7
Sub-Saharan Africa, deserts 67–8
subsidies 72–4; *see also* feed-in tariffs
Sunrise Powerlink, California 41
sunshine *see* insolation
support 74–81; *see also* subsidies
Sweden **15**

targets, solar in India 137–9
tax relief 84–5

technology overview 20–2
tellurium 49, 51, 147
Terrasol solar power project in Spain 122
Thar Desert 66, 121, 140, 153
thermal cycle 23–5
thin-film technologies 47–51, 112, 147, 156–7; *see also* cadmium telluride (CdTe)
thorium 12–13
towers *see* power towers
tracking systems 31, 33, 53, 54
Trans-Green initiative 135
Tunisia 135
Turkey 77, 103

UK 1, 2, 6, 9, 13; feed-in tariffs 77; renewable obligation 75; shale gas 11; wind power 143
United Arab Emirates 92; *see also* Masdar City, Abu Dhabi; MENA (Middle East and North Africa)
uranium 12–13
USA 3, 17, 84–6, 87, 151–2; CSP plants 97–104, **102**; deserts 62–3; emissions **15**, 16; large scale PV 115–17; *see also* California; Hawaii

water 4–5, **29**, 148–51; cooling 28–30, 41, 53, 107; *see also* desalination
WEEE (Waste Electrical and Electronic Equipment) 147
wind power 16, 76, 129–30, 143, 153; costs 72; Spain 45